CRITIQUE OF RADIOMETRIC DATING

by

HAROLD S. SLUSHER, M.S., D.Sc.

Assistant Professor of Physics
University of Texas at El Paso
and
Research Associate
Institute for Creation Research
San Diego, California

ICR Technical Monograph No. 2

Institute for Creation Research
San Diego, CA 92116

CRITIQUE OF RADIOMETRIC DATING

Institute for Creation Research
2716 Madison Avenue
San Diego, CA 92116

Library of Congress Catalog Card
Number 81-80259
ISBN 0-932766-04-8

Cataloging in Publication Data

Slusher, Harold Schultz, 1934-
Critique of radiometric dating. 2d Ed.
1. Radioactive dating. I. Title.
II. Series: Institute for Creation Research.
ICR technical monograph, no. 2, rev.

ISBN-0-932766-04-8 539.77 81-80259

Printed in United States of America.

THE AUTHOR

Professor Harold S. Slusher earned his B.A. from the University of Tennessee (mathematics and physics) and his M.S. from the University of Oklahoma (physics and astronomy). In 1975 he was awarded an honorary D.Sc. degree by Indiana Christian University in recognition of his work on the time scale for the cosmos. For many years he served as Director of the University of Texas at El Paso Kidd Memorial Seismic Observatory. He is Assistant Professor on the faculty of the Department of Physics at the University of Texas at El Paso and also serves as Research Associate in Geoscience and Astronomy with the Institute for Creation Research and is Adjunct Chairman of the Department of Physical Sciences at Christian Heritage College.

Professor Slusher's fields of interest and research include solar system astronomy, terrestrial heat flow, and cosmogony. In addition to this monograph *Critique of Radiometric Dating,* he has also written other Institute for Creation Research monographs: *Age of the Cosmos; The Origin of the Universe: An Examination of the Big Bang and Steady-State Cosmogonies; The Age of the Solar System: A Study of the Poynting-Robertson Effect and Extinction of Interplanetary Dust;* and *The Age of the Earth: A Study of the Cooling of the Earth Under the Influence of Radioactive Heat Sources.*

ACKNOWLEDGMENTS

This work is an examination of the major radiometric dating methods. It involves a study of the principles, assumptions, and methods of the most-used radioactive "clocks." Discussions of various aspects of this study with Dr. Duane T. Gish were of much value. I owe a great debt to Dr. Melvin A. Cook whose very thorough works regarding radiometric "clocks" were invaluable.

I would like to dedicate this work to my mother and my late father. They have always been a source of inspiration and encouragement.

TABLE OF CONTENTS

Chapter 1
THE AGE OF THE EARTH AND ITS DETERMINATION

An investigation of the age of the earth and the times of occurrence of various geological events is essential in any study of the physical history of the earth. The evolutionist needs vast spans of time in the history of the earth. His hypothesis regarding the origin and development of life on the earth says that almost imperceptible changes have occurred very gradually over vast time intervals to produce the world that we see before us. This concept must feed on time. The importance of the study of the chronometry of physical events on the earth and in the universe in arriving at the truth about the past cannot be overemphasized.

Time can be sensed only by the events that occur within its span. But, in the study of the earth, we never observe past events themselves, only the effects of these events on the rocks of the earth. The student of the earth tries to infer from a study of the rocks the events that gave rise to the various effects observed. In all studies of the past, but particularly in geochronology, it is tremendously important to keep actual observations separate from the speculative inferences. It is very easy for the two to become interlaced in the investigation of such a complex problem as geochronology, but they are very different. Facts are true, but inferences are derived and may be either true or false or only partially true.

In much of the work done by geochronologists it is not always easy to separate fact from hypothesis because they fail many times to make clear where fact ends and supposition begins. It seems that scientists oftentimes try to force science to express their philosophy as to what the world was like in the past.

The geologist says that essentially the past should be interpreted in the light of the present—"the present is the key to the past." Though the study of these present processes has continued at an ever-increasing pace for decades, we are still appallingly ignorant of many of the simplest details of what is happening today, such as ocean currents, sedimentation, causes of ice sheets, behavior of the subcrust, laws governing the flow of streams, etc. Furthermore, many data around the earth indicate that the rates of the processes operating in the past have been radically different from those of the present. Catastrophic happenings in the past may have radically altered the distribution of radioactive minerals and their decay products. Volcanism, for example, which has obviously occurred in the past on a large scale, would radically alter the carbon-14 to carbon-12 ratio (C^{14}/C^{12}) in the atmosphere, thus affecting the C^{14} "clock." "The present is the key to the past" statement, if referring to rates of activity, certainly has no scientific foundation.

The determination of the times of occurrences of the various geological events depends first on the determination of rates, such as the rate of decay of uranium and other radioactive elements into their daughter products, influx rates of salts into the oceans, etc. These rates may have been very different in the past from what they were when the measurements were actually made. Secondly, the determination depends on the initial conditions, which cannot be ascertained in any direct way, e.g., the initial ratios of parent-to-daughter elements of radioactive series. The third factor involves hypotheses about the origin of the crust of the earth and the physical/chemical processes that have taken place and are presently occurring.[1]

The highly speculative nature of geochronology seems rather obvious in the light of the many assumptions involved above. When one considers the dogmatic way its questionable findings are used, he is appalled if not indignant of the apparent intellectual dishonesty involved.

The second law of thermodynamics says that all natural processes are deteriorative or degenerative. Natural processes are changing the universe in a way similar to the unwinding of a clock spring that loses organization by the ticking of the clock. An outside agent must be present and active to make the "clock" and to "wind" it up to begin with. It is not possible to work backwards in a situation where there is a disordering effect continually taking place and arrive at a unique description of past conditions. The scientific method is not applicable when working back into the past where there were no observations.

The age of the earth has had nearly as many values as the number of people who have studied the matter. Over the years the evolutionist has pushed the age further and further back into a remote dim past. He has used various techniques to get these alleged ages. In recent years, however, creationists who have studied these techniques have found serious flaws in them and have made very significant corrections on the techniques, corrections that seem to be demanded by the data. These corrections have led to young ages for geological events by the same radioactive dating methods the uniformitarians have claimed give long ages.

There have been many attempts to determine the age of the earth during the past. The maximum age has been calculated by Helmholtz from the rate of contraction of the sun, by Kelvin from heat flow at the surface of the earth, by Joly from the transport of salt to the oceans, by Holmes from certain radioactivity measurements, and by recent investigators using a wide range of radiometric techniques. Most of these investigators have been guided by evolutionist presuppositions and assumptions. Theories regarding the mechanism of evolution calling for greater and greater time have led to a time scale of great elasticity (no time span is too great—the more time the better). Fred Whipple has pointed out that on the average the "age" of the earth has been doubling every 15 years for the past three centuries.[2] Currently, the evolutionists are maintaining with dead certainty that the average "age" of the earth from various radiometric techniques is 4.6 billion years (within a few hundred million years). After all the claims they have made in the past which have turned out wrong, they still do this with a straight face.

Most creationists, on the other hand, have viewed the evidence regarding the age of the earth as pointing to a very young age of from about 7,000 years to 10,000 years. They maintain that not only do various physical indicators, not radiometric in nature, but also many of the radiometric schemes themselves, when properly interpreted, point strongly to a young age.

Chapter 2

CRITERIA FOR USABLE GEOCHRONOMETERS

Geologic clocks work on the assumption that some physical quantity is being produced at a constant rate. The clock can give the elapsed time if we know how much of the physical quantity was present when the clock started (Q_O), its rate of production (R), and the present amount of the quantity (Q). Looking at this in a simplified form, all we would need to do to find the geologic time (t) by this clock is to divide the amount of the physical quantity produced (ΔQ) during some time interval (t) by its constant rate (R) of production of, $t = \frac{\Delta Q}{R} = \frac{Q - Q_O}{R}$.

Various clocks are being used today to tell geologic time. We will examine later some of the radiometric time systems to see if they are valid clocks, but for the moment let's take a look at the criteria for a valid clock.

What characteristics must a "clock" have to be a valid indicator of geologic time?

A. There must be a change in some physical quantity with time, and the physical changes that are involved in the clock must be such that they can be measured in an understandable way.

B. The clock must have an accuracy and sensitivity that is sufficiently good to measure the time intervals in question.

C. How "tightly" was the clock wound to begin with? Was there an initial amount of the physical quantity present when the clock started?

D. The clock has to run at a constant rate. If it runs fast part of the time and slow at other times, we will need information regarding this so we can allow for a changing rate. This actually seems impossible to obtain since no one can go back into the geologic past to find this information. Furthermore, we need to know whether the clock has been running all the time or just part of the time. As will be seen later in this monograph, the mode of transformation of elements into daughter elements plays a major role in the nature of the rate, whether variable or invariable.

E. The clock must not have been "reset" in any way. For example, all of the so-called "radiogenic" Pb (lead) may not have been produced by radioactive decay, but neutron reactions may also have occurred to produce lead, thus moving the "hands" forward, resetting the clock.

These requirements must be stringently met if the physical process is to be used as a clock at all. Let us now examine some of the radiometric methods to see if they fulfill these criteria.

Chapter 3

RADIOMETRIC DATING METHODS

Arthur Holmes has said, "Until the discovery of radioactivity, geologists were in the same position as a historian who knew, for example, that the Roman invasion of Britain was followed by the Norman Conquest and that both these events occurred before the Battle of Waterloo, but who could not find any record of the dates of these or any of the other great events of history."[3] He argued that up to the discovery of radioactivity the geologist had only established a relative chronology, "a chronology without years."[4] He emphasized that radioactivity provides a way of establishing an absolute chronology.

In principle, radioactive dating works similarly to an hourglass for telling time (see Fig. 1). If we know that it takes 60 minutes for the sand in the top chamber of the hourglass to reach the bottom of the glass, then we know that on the average 1/3600 (since there are 3600 seconds in an hour) of the sand falls from one chamber to the other in one second. To get the approximate number of seconds that had elapsed, or the time since the sand had started falling during the hour, we could measure the amount of sand which had fallen up to that time and divide it by the amount that falls in a second.

In the early days of radioactive dating methods, radioactive disintegration of certain elements was used in a similar manner to determine the age of the earth and date various geological events. For example, in the case of the isotope of uranium whose atomic weight is 238, 4.9×10^{-18} of whatever mass of uranium is present decomposes into thorium of atomic weight 234 each second. The thorium then disintegrates into protoactinium, and so on down the series through radium and radon, eventually becoming lead of atomic weight 206. Approximately, then, the age was taken as:

$$\frac{\text{amount of } Pb^{206} \text{ produced by decay of } U^{238}}{(\text{fractional rate of conversion of } U^{238} \text{ to } Pb^{206})(\text{amount of } U^{238} \text{ present now})}$$

$= Pb^{206}/\lambda_{238}U^{238}$.

Now, however, the geochronologist uses a more accurate formula than this one. It would be to our advantage to look briefly into the physics of radiometric methods.

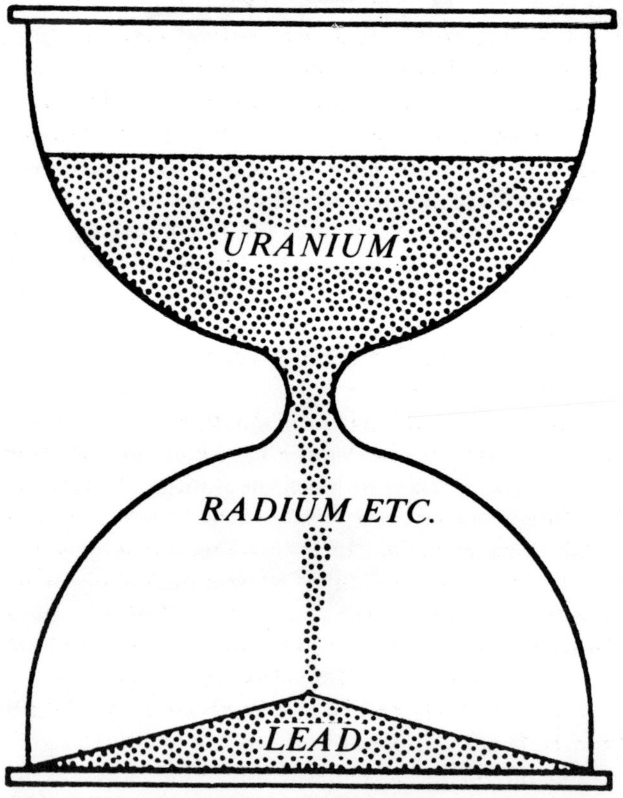

Figure 1. *Radioactive disintegration tells time like an hourglass.*

A. PHYSICS OF RADIOMETRIC DATING METHODS

1. Atomic Structure

Matter is composed of atoms which combine to form molecules, and often these are arranged into crystals. Every atom has a central nucleus about $^{-12}$ to 10^{-13} centimeter in diameter and which contains protons and (ordinarily) neutrons. Protons are particles which carry a positive electrical charge and have a mass of 1.67×10^{-24} gm. A neutron is electrically neutral and its mass is slightly greater than that of the proton. According to Henry Faul, nuclear matter is so dense that an imaginary nucleus one foot in diameter would weigh as much as 10 cubic miles of rock, much more than all the trucks and freight cars in the world fully loaded.[5] The nucleus is surrounded by a spherical cloud of electrons with a diameter of about 10^{-8}cm. Henry Faul points out that in our imaginary nucleus its "outermost electrons would be more than a mile away."[6] The electron is a very small particle (about 1/1840th the mass of the proton).

Since a neutron and a proton have about the same mass, the number of these particles (protons and neutrons) inside a nucleus gives the mass of the nucleus in terms of the mass of a neutron. This number is referred to as the atom's atomic mass. The number of protons inside the nucleus of the atom is the atomic number of that element, and this is the same as the number of electrons in the electron cloud of the atom if it is electrically neutral (no balanced electrical charge). If the atom has lost or gained electrons, it is positively or negatively charged, respectively, and is said to be ionized.

A given species of an atom is called a nuclide. Nuclides of the same atomic mass, A, but of a different atomic numbers are called isobars. Nuclides that have the same number of protons but a different number of neutrons are called isotopes of that particular element. Since they have the same atomic number and thus the same number of electrons, all the isotopes for any one element have like chemical properties. Isotopes differ only in their atomic masses.

When we write $_ZE^A$, we mean that the element E has Z protons, A protons and neutrons, and (A-Z) neutrons inside the nucleus. Thus $_{92}U^{238}$ specifies that this particular nucleus of a uranium atom contains 92 protons and 146 neutrons, and that its atomic mass is 238 (=92 + 146).

An alpha-particle (α-particle) is practically identical to a helium nucleus and is denoted by the symbol $_2He^4$, meaning that it is composed of 2 neutrons and 2 protons. The beta-particle (β-particle), denoted by $_{-1}e^0$, has a very small mass compared to a proton and carries the same electrical charge as an electron. A gamma-ray

(ϒ-ray) is electromagnetic radiation similar to x-rays but of a higher frequency.

2. Radioactivity

Radioactivity is the term applied to those natural events in which a nucleus of an atom becomes unstable and emits or "throws out" an α- or a β-particle. Such an event is ordinarily accompanied by the emission of various radiations.

A radioactive nucleus may decay in different ways. The nucleus may emit an α-particle at a high velocity. In alpha decay, the atomic number, Z, of the nucleus decreases by two because of the removal of two protons, and the mass number, A, decreases by four. The kinetic energy (energy of motion) of alpha particles is usually several million electron volts (Mev). One Mev is the kinetic energy acquired by an electron in falling through a potential difference of one million volts.

Another way of radioactive decay is beta emission. Probably this type of decay involves the decay of a neutron into a proton and an electron, with the emission of the electron as the beta particle. The number of protons in the beta-decaying nucleus increases by one, and the parent nucleus becomes the nucleus of the next higher element in the periodic table. The energy of the beta particle emitted by the nucleus may range from zero to the maximum available for the transition.

Decay may occur by electron capture. The nucleus captures an electron from the innermost shell of the electron cloud, and the electron combines with a proton, changing it into a neutron. Electron capture reduces the atomic number by one, leaving the atomic mass constant. The nucleus produced by electron capture may be in an excited state and may then decay to a state of lower energy by emitting the excess energy as a gamma ray.

As a result of these actions, the nucleus loses its original characteristics and becomes the nucleus of a different element. Radioactivity occurs naturally in all elements having an atomic number greater than 80; though, in some rare instances, it also occurs in elements of lower atomic number —potassium (with atomic number 19) is radioactive.

Radium, which is a very dense metal, emits α-particles and slowly changes into a new element which appears in its place. If you bought a pound of radium, your descendants would find that in about 1600 years there would only be about half of it left. The interval of time that it takes for one-half of the quantity of a radioactive substance to decay is called its half-life. The decay constant is the proportion of the substance that decays in a unit time. Unlike linear rates of depletion common in most everyday processes, radioactive depletion is exponential. This is illustrated in Fig. 2. (For a list of some of the long-lived radioactive nuclides see Table 1.)

Thus, some chemical elements change spontaneously to others by natural transmutation. In addition to these spontaneous transmutations, the particles emitted by these elements may cause transmutations of the nuclei of other elements in their vicinity. Also, by bombarding stable atoms of almost any element with high-energy particles, new radioactive elements can be made, each with a family tree of radioactive nuclides ending up as some stable atom.

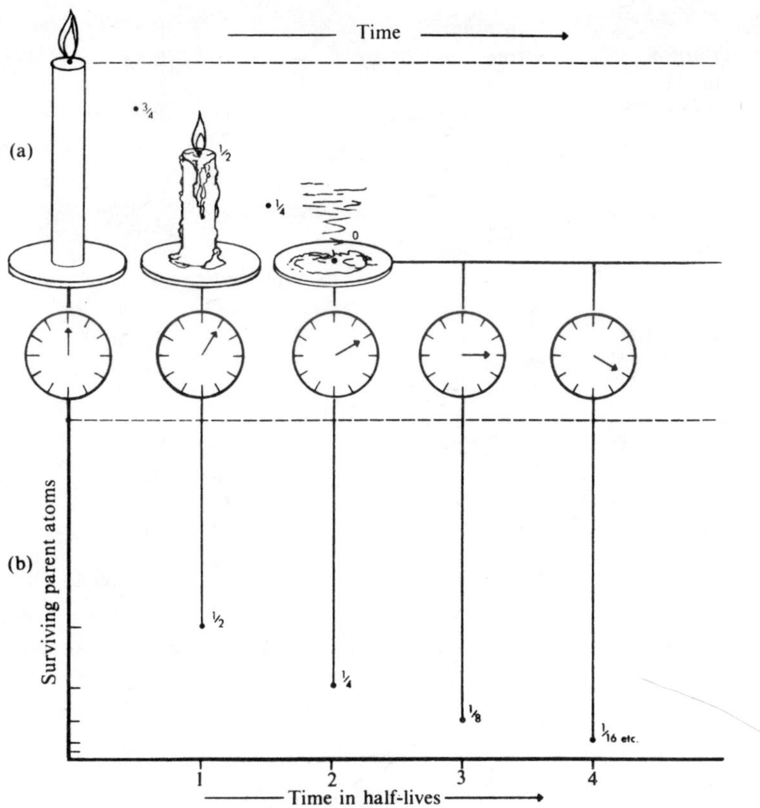

Figure 2. *(a) Uniform straight line depletion of most everyday processes. (b) By contrast, the radioactive decay curve approaches zero line asymptotically. The end of one half-life interval is the beginning of a new one.*

Table 1
Long-lived Radioactive Nuclides

Parent	Daughter	Half-life	Type of Decay
Potassium-40	Argon-40	1.3×10^{9} total	Electron capture
	Calcium-40		Beta
Vanadium-50	Titanium-50	$\sim 6 \times 10^{15}$ total	Electron capture
	Chronium-50		Beta
Rubidium-87	Strontium-87	4.7×10^{10}	Beta
Indium-115	Tin-115	5×10^{14}	Beta
Tellurium-123	Antimony-123	1.2×10^{13}	Electron capture
Lanthanum-138	Barium-138	1.1×10^{11} total	Electron capture
	Cerium-138		Beta
Cerium-142	Barium-138	5×10^{15}	Alpha
Neodymium-144	Cerium-140	2.4×10^{15}	Alpha
Samarium-147	Neodymium-143	1.06×10^{11}	Alpha
Samarium-148	Neodymium-141	1.2×10^{13}	Alpha
Samarium-149	Neodymium-145	$\sim 4 \times 10^{14}$?	Alpha
Gadolinium-152	Samarium-148	1.1×10^{11}	Alpha
Dysprosium-150	Gadolinium-152	2×10^{14}	Alpha
Hafnium-174	Yeterbium-170	4.3×10^{15}	Alpha
Lutetium-176	Hafnium-176	2.2×10^{10}	Beta
Rhenium-187	Osmium-187	4×10^{10}	Beta
Platinum-190	Osmium-186	7×10^{11}	Alpha
Lead-204	Mercury-200	1.4×10^{17}	Alpha
Thorium-232	Lead-208	1.41×10^{10}	6 Alpha + 4 Beta
Uranium-235	Lead-207	7.13×10^{8}	7 Alpha + 4 Beta
Uranium-238	Lead-206	4.51×10^{9}	8 Alpha + 6 Beta

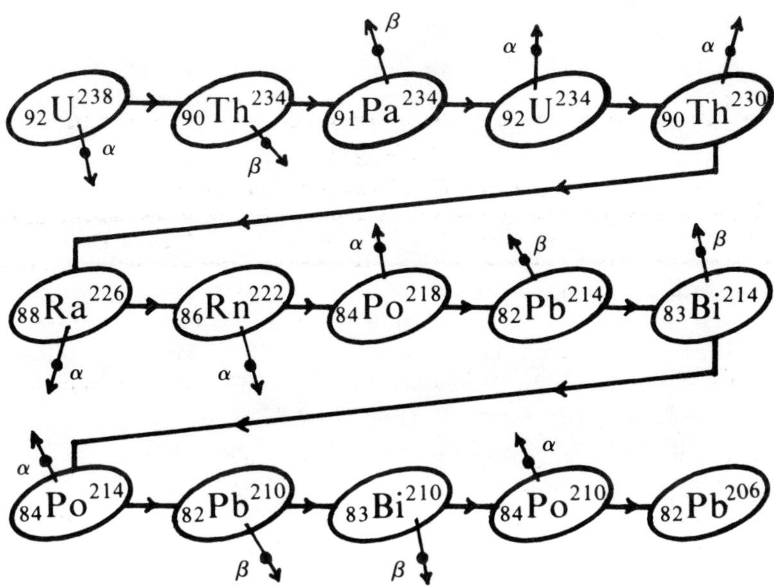

Figure 3 (a). *The uranium-radium series of radioactive changes. The subscript is the number of protons in the nucleus, or the atomic number Z. The superscript equals the sum of the neutrons and protons in the nucleus, usually called the mass number A. Emission of an α-particle $_2He^4$, since it contains two protons and two neutrons, decreases A by 4 and Z by 2. When a β-particle $_{-1}e^0$ is ejected, A remains constant and Z increases by 1.*

Figure 3 (b). *Decay Scheme of potassium-40. Electron capture (EC) leads to an excited state of argon-40, which decays to ground state by gamma ray emission.*

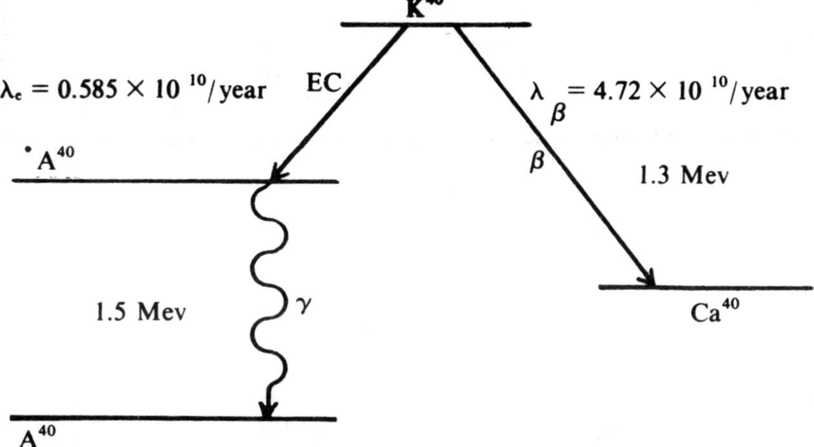

3. The Radioactive Series

The radioactive elements can be arranged in groups or series in such a manner that any element in the group is formed by the disintegration of the element ahead of it (see Fig. 3). There are a few such groups that occur naturally, and there are groups that can be produced by bombardment of atoms with high-speed neutrons, protons, etc. Some of the series that occur naturally that are of major interest in radiometric dating are: (1) the uranium-radium series, starting with U^{238} and ending with Pb^{206}; (2) the actinium series, starting with U^{235} and ending with Pb^{207}; (3) the thorium series, starting with Th^{232} and ending with Pb^{208}; (4) the rubidium-strontium series, starting with Rb^{87} and ending with Sr^{87}; (5) the potassium-argon series, starting with K^{40} and ending with Ar^{40}; (6) the carbon-14 series, starting with C^{14} and ending with N^{14}.

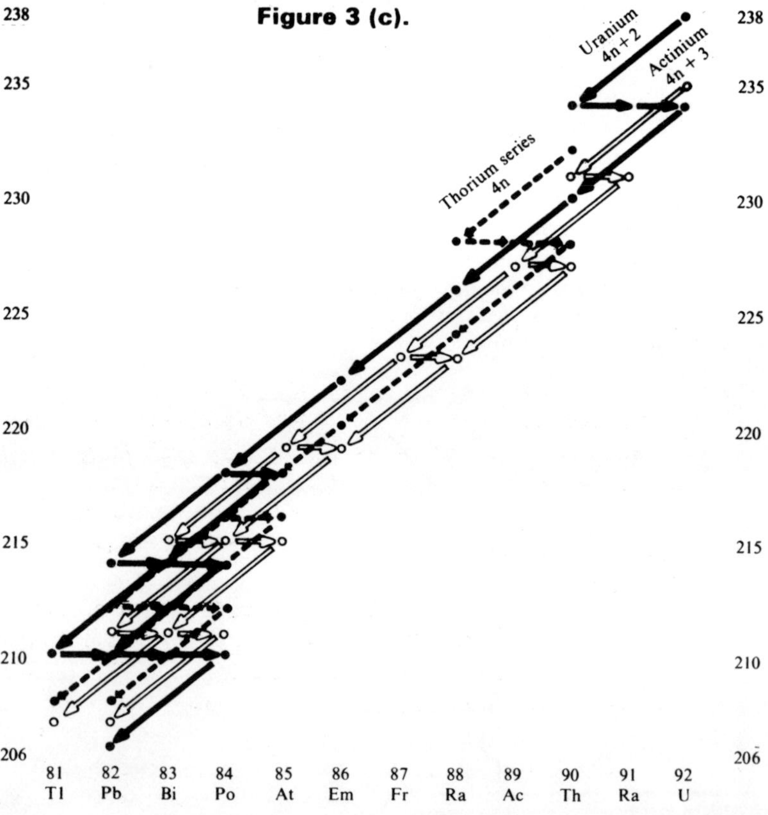

Figure 3 (c).

4. Radioactive Decay According to the Conventional Theory

Potassium-40, through radioactive decay, is converted into argon-40. How long does it take for these changes to occur? How long does it take for a certain quantity of potassium to change into argon? Does it change at a uniform rate? Let's consider the answers to these questions as they apply to radioactive elements.

There is no way of predicting when a specific atom of a radioactive element will disintegrate. It may eject an α-particle or a β-particle within the next few seconds, or it may remain stable for a very long time. If you prepared to watch a few radioactive atoms you might not have time to get settled before one decayed, or you might be able to take a very long nap (maybe millions of years) before anything happened. However, if there are a large number of radioactive atoms of an element, the picture is completely changed, for there would be decays taking place all the time.

If $T_{1/2}$ is the half-life of a radioactive element and N_O is the number of atoms initially (at t = 0 seconds), there would be $N_O/2$ number of atoms remaining after $T_{1/2}$ seconds have elapsed. Of these $N_O/2$ number of atoms, half of them would disintegrate during the next $T_{1/2}$ seconds. No matter how many atoms are present at any specified time, one-half of them will have decayed $T_{1/2}$ seconds later. The decay constant, λ, of an element is the fractional part of the atoms that disintegrate each second. The activity of a radioactive sample is a measure of the number of disintegrations per second the sample undergoes. When the rate of production of atoms of a sample equals the rate of decay of the atoms of this sample, equilibrium exists for the members of the radioactive series.

Let's develop a mathematical expression, using a rather elementary approach, that describes radioactive decay. Consider the radioactive decay as taking place over an interval of time t and say that t can be divided into n equal short time intervals [$\Delta t = (t/n)$]. N_O is the number of atoms of the element present at the beginning of the time t and N will be the number of atoms left at the end of the time t. Let λ represent the fraction of the atoms decaying per unit of time.

In the short time interval Δt, $\lambda \Delta t$ times N_O atoms will decay. The number of atoms left will be the original total minus those lost by decay, $N_O - N_O \lambda \Delta t$, which can be expressed as $N_O (1 - \lambda \Delta t)$.

In the second short time interval Δt the atoms again will decay in the same (supposed) constant proportion. Thus, $N_O - N_O \lambda \Delta t$ is multiplied by $\lambda \Delta t$ to find the number of atoms which have decayed. This gives $N_O \lambda \Delta t - N_O \lambda^2 \Delta t^2$. Subtracting this from the number of atoms that were left at the beginning of the second short time interval we find the number left at the end of the second interval:

$$N_O - N_O\lambda\Delta t - (N_O\lambda\Delta t - N_O\lambda^2\Delta^2)$$
$$= N_O - N_O\lambda\Delta t - N_O\lambda\Delta t + N_O\lambda^2\Delta t^2$$
$$= N_O - 2N_O\lambda\Delta t + N_O\lambda^2\Delta t^2$$
$$= N_O\ (\ 1 - 2\ \lambda\Delta t + \lambda^2\Delta t^2)$$

which is $N_o\ (1 - \lambda\Delta t)^2$, the number of atoms left at the end of the second Δt interval. We could go on through a third Δt, getting $N_O \cdot (1 - \lambda\Delta t)^3$, and a fourth, $N_O\ (1 - \lambda\Delta t)^4$, and so on. Let us consider n intervals of Δt. After n intervals of Δt, the number of atoms left will be $N_O\ (1 - \lambda\Delta t)^n$.

Now, let there be $n\Delta t$ intervals of time in the total time t: so, $n\Delta t = t$ or $\Delta t = t/n$. Substitute t/n for Δt in the general expression of the number of remaining atoms. Thus, $N_O\ (1 - \lambda\Delta t)^n$ becomes $N_O\ (1 - \lambda t/n)^n$. This leads to an equation that gives the law of decay, stating the number of remaining atoms N at any time in the sample. Thus, $N = N_O\ (1 - \lambda t/n)^n$.

To get a better form for this equation, we shall have to make the Δt intervals very small. The smaller they become, the larger n, the number of these intervals, must become. The greatest accuracy comes when we let Δ t approach zero; then n approaches infinity.

There is a number e (=2.718) which has the property such that e^{χ} is equal to the limit of $(1 + \chi/n)^n$ when n goes toward infinity. For our case then, $\lim (1 - \lambda t/n)^n = e^{-\lambda t}$. Using this exponential form, we can write the law of decay as $N = N_O e^{-\lambda t}$. If N_O is the original number of, say, uranium atoms in a sample formed some time ago and λ is the proportion of the U atoms disintegrating per second, then N will be the number of atoms of uranium left at the end of time t.

A more rigorous approach to the problem follows. Suppose we start with A_O atoms of a radioactive element whose disintegration constant is λ_1. Let A be the number of atoms of the parent element present at any time t thereafter. The rate of disintegration will be

$$\frac{dA}{dt} = -\lambda_1 A \qquad \text{(a)}$$

Writing this in the form

$$\frac{dA}{A} = -\lambda_1 dt \qquad \text{(b)}$$

It can be seen that the solution will be of the form

$$\ln A = -\lambda_1 t + K \qquad \text{(c)}$$

where ln stands for logarithm to the base e and K is a constant which can be determined from the boundary conditions shown below. Throughout, ln will refer to natural logarithms to the base e and log will refer to logarithms to the base 10.

Equation (c) can be written

$$A = e^{(-\lambda_1 t + K)} \tag{d}$$

At time $t = 0$, $A = A_O$, so

$$A_O = e^K \tag{e}$$

and

$$A = A_O e^{-\lambda_1 t} \tag{f}$$

The number of new atoms which have been formed from A to any time t is

$$B = A_O - A = A_O(1 - e^{-\lambda_1 t}) = A(e^{\lambda_1 t} - 1) \tag{g}$$

This is the number of atoms of element B of the series which is present at time t only if B is not radioactive. If B is radioactive then,

$$\frac{dB}{dt} = \lambda_1 A - \lambda_2 B \tag{h}$$

It can be shown (by substituting in (h)) that the solution of this equation is

$$B = \frac{A_O\lambda_1}{\lambda_2 - \lambda_1}(e^{-\lambda_1 t} - e^{-\lambda_2 t}) = \frac{A\lambda_1}{\lambda_2 - \lambda_1}\left[1 - e^{(\lambda_1 - \lambda_2)t}\right] \tag{i}$$

In a similar manner, if the third element of the series C is radioactive

$$\frac{dC}{dt} = \lambda_2 B - \lambda_3 C \tag{j}$$

$$C = A_O\lambda_1\lambda_2\left[\frac{e^{-\lambda_1 t}}{(\lambda_3 - \lambda_1)(\lambda_2 - \lambda_1)} + \frac{e^{-\lambda_2 t}}{(\lambda_1 - \lambda_2)(\lambda_3 - \lambda_2)} + \frac{e^{-\lambda_3 t}}{(\lambda_1 - \lambda_3)(\lambda_2 - \lambda_3)}\right] =$$

$$A\lambda_1\lambda_2\left[\frac{1}{(\lambda_3 - \lambda_1)(\lambda_2 - \lambda_1)} + \frac{e^{(\lambda_1 - \lambda_2)t}}{(\lambda_1 - \lambda_2)(\lambda_3 - \lambda_2)} + \frac{e^{(\lambda_1 - \lambda_3)t}}{(\lambda_1 - \lambda_3)(\lambda_2 - \lambda_3)}\right] \tag{k}$$

No matter how long the series, the form of the differential equation will be

$$\frac{dN}{dt} = \lambda_m M - \lambda_n N \tag{l}$$

where M and N are the numbers of atoms of two successive members of the series present at any time t, and λ_m and λ_n are their decay constants. The number of atoms of N will be

$$N = A_O(F_1e^{-\lambda_1 t} + F_2e^{-\lambda_2 t} + \ldots + F_n e^{-\lambda_n t})$$

$$= A\ F_1 + F_2 e^{(\lambda_1 - \lambda_2)t} + \ldots + F_n e^{(\lambda_1 - \lambda_n)t} \qquad \text{(m)}$$

$F_1, F_2, F_3, \ldots, F_n$ are the constant coefficients of the $e{-}\lambda^{it}$ terms in equation (k). These equations can be simplified if λ_1 is much smaller than any of the other decay constants in the series. In this case, after a time long enough that $\lambda_2 t$ is much greater than unity, (i) becomes

$$B \approx \frac{\lambda_1 A}{\lambda_2} \qquad \text{(n)}$$

Similarly, (k) will become

$$C \approx \frac{\lambda_1 A}{\lambda_3} \qquad \text{(o)}$$

and by analogy, for any element N except the end-product element,

$$N \approx \frac{\lambda_1 A}{\lambda_n} \qquad \text{(p)}$$

The end-product element P is not radioactive. Therefore, λ_p is 0, which is not small compared with λ_1.

The meaning of the approximation, $\lambda_n t$ is much greater than 1, can be seen if we note the relation of λ_n to the corresponding half-life $T_{1/2}$ of an element. At $t = T_{1/2}$, (f) becomes

$$A = A_O e^{-\lambda_1 T_{1/2}} \qquad \text{(q)}$$

Solving this for $T_{1/2}$,

$$T_{1/2} = \frac{\ln 0.5}{-\lambda_1} = 0.69315\lambda_1^{-1} \qquad \text{(r)}$$

This relation between $T_{1/2}$ and the disintegration constant applies to any radioactive element. Therefore, the approximation $\lambda_n t$ is much greater than 1 means that $0.69315t/T_n$ greatly exceeds 1, or that many half-lives of the element under consideration have elapsed. Since the half-lives of all the intermediate elements in the uranium and thorium series are small compared with the half-lives of their parents, this is usually a good approximation.

To find the amount of the end product present, it is sufficient to note that the total number of atoms present at any time (not counting the helium nuclei formed) must always equal the number present originally:

$$A_O = A + B + C + \ldots + N + P. \qquad \text{(s)}$$

Solving this for P and using equations (f) and (p) to simplify the result,

$$P = A_O - A - B - C - \ldots - N$$

$$\approx Ae^{\lambda_1 t} - A - \frac{\lambda_1 A}{\lambda_2} = \frac{\lambda_1 A}{\lambda_3} - \ldots - \frac{\lambda_1 A}{\lambda_n}$$

$$= A(e^{\lambda_1 t} - 1 - \lambda) \qquad \text{(t)}$$

where λ is the fraction of the elements in intermediate stages of disintegration. After sufficient time has elapsed, λ approaches the value

$$\lambda = \sum_{m=2}^{m=n} \frac{\lambda_1}{\lambda_m}$$

which is a very small quantity for the uranium and thorium series ($\lambda \approx 0.74 \times 10^{-4}$ for the U^{238} series) or $P = A(e^{\lambda_1 t} - 1)$.

Equation (t) can be solved for t, giving

$$t = \lambda_1^{-1} \ln(1 + \frac{P}{A} + \lambda) \quad . \qquad \text{(v)}$$

If $(P/A) + \lambda$ is less than 1, the log term can be expanded as an infinite series

$$t = \lambda_1^{-1}\left[(\frac{P}{A} + \lambda) - 1/2(\frac{P}{A} + \lambda)^2 + 1/3(\frac{P}{A} + \lambda)^3 \ldots\right] \qquad \text{(w)}$$

When P/A is much less than 1 but much larger than λ, the time t is approximately

$$t \approx \frac{P}{\lambda_1 A} \cdot \qquad \text{(x)}$$

Thus, the decay equations are: $N(Pb^{206}) = N(U^{238})(e^{\lambda 238 t} - 1)$;

$N(Pb^{207}) = N(U^{235})(e^{\lambda 235 t} - 1)$; $N(Pb^{208}) = N(Th^{232})(e^{\lambda 232 t} - 1)$;

and $$\frac{N(Pb^{207})}{N(Pb^{207})} = \frac{N(U^{235})}{N(U^{238})} \quad \frac{(e^{\lambda 235 t} - 1)}{(e^{\lambda 238 t} - 1)} \cdot$$

N represents the number of atoms present now.

If D* represents the daughter isotope produced by radioactive decay, the number of atoms of the daughter isotope formed by radioactive decay at some time t will be $D^* = N_O - N$ (= the difference of the initial number of parent atoms and the number of parent atoms existing at time t). This may be rewritten as $D^* = N_O - N_O e^{-\lambda t}$ $= N_O(1 - e^{-\lambda t})$. As an example, take the case of U^{238} decaying to Pb^{206} or Th^{232} decaying to Pb^{208}. If there were U_O^{238}, Pb_O^{206}, Th_O^{232}, and Pb_o^{208} present when the decay started, at the present time t the amount of lead found in the sample, assuming no differentiation and contamination, will be:

$$Pb^{206}_{now} = Pb_o^{206} + U_o^{238} (1 - e^{-\lambda 238 t}) \qquad (y)$$

or

$$Pb^{208}_{now} = Pb_o^{208} + Th_o^{232} (- e^{-\lambda 232 t}). \qquad (z)$$

Suppose somehow that U^{238} and Pb^{206} are known; then, these equations can be solved to obtain the time t since the rock in which the lead and uranium is found was mineralized. Similar equations can be set up for the various radioactive series used in radiometric dating. Briefly, then, this is the conventional way a radiometric age is found.

5. Dudley's Hypothesis Regarding Radioactivity

The conventional explanation of radioactivity maintains that radioactivity results from "spontaneous" processes by which certain atoms emit various particles and gamma rays. H.C. Dudley [7] points out that the use of the terms "spontaneous" and "causality" is illogical. He proposes that radioactivity is not a spontaneous process whereby, without cause, atoms could transmute themselves to an entirely new species.

Dudley maintains that the experimental apparatus used in the early studies of radioactivity had inherent limitations that led to the following erroneous conclusions[8]:

(a) that radioactive decay rates are constant;

(b) that these rates cannot be altered by change of the energy state of the electrons orbiting the nucleus;

(c) that radioactivity results from processes which involve only the atomic nucleus.

From 1949 through 1972 rather easily induced changes in the disintegration rates of 14 radionuclides were produced using changes in pressure, temperature, chemical state, electric potential, stress of monomolecular layers, etc. These experimental observations along with the development of complex and alternative disintegration modes has led Dudley and others to the conclusion that radioactivity must result from some complex cause. Dudley considers the decay "constant" to be a variable. He contends that the value is dependent on the energy state of the entire atom as the basic unit system, not just on the energy state of the atom's nucleus.

Some examples of the experiments and the results referred to by Dudley are the following: Segre, Wiegand, and Leiniger, utilizing ^{7}Be (a radionuclide decaying by capture of a K-shell electron), experimentally induced a change in the radioactive decay rate. It was thought that the electron density near the nucleus would be altered by variation in the energy state of the two L-shell valence electrons produced by chemical combinations. A change of the decay rate of about 0.1% was found in Be/BeO.

The decay rate ^{90}Nb has been determined in combination with fluoride. When it was compared with the elemental form, the half-life was changed by 3.6%. Dudley believes the decay is caused by internal conversion and is influenced by variations of the energy content of the four N-shell valence electrons. This alters the effective charge density adjacent to the nucleus.

Anderson experimentally demonstrated an induced variation of decay characteristics of $\beta-$ emitters. He observed small but non-random changes in the pattern of observed counts from ^{14}C. ^{14}C was incorporated into a polar organic molecule and the bonds as a monomolecular layer was stressed.

Small changes in the energy content of L-shell electrons of ^{7}Be, and like changes in the 0-shell electrons of ^{137}Cs, have caused variations in decay charactertistics of 12 other radionuclides.

Dudley summarizes his hypothesis: "The equation $N=N_Oe^{-kt}$ can no longer be considered valid. The decay "constant" has been shown to be a variable, dependent on the energy content of the entire atom, rather than being dependent only on the mass-energy relationship of the nucleus. Thus k becomes a stability index, defining the state of certain atoms which have been termed linear resonant systems; subject to parametric excitation."[9]

Since in the conventional theory radioactive decay is described as a series of truly random, unrelated events, each occurring "spontaneously," without any prior cause, the "disintegration constant by definition "becomes a true constant that cannot be altered." But the experimental evidence indicates it is possible to alter radioactive decay rates and thus it would seem logical to conclude that such processes result from some finite, causal process.

De Broglie has proposed space to be filled with a "subquantic medium" or "a gas made up of leptons and probably neutrinos." Some astrophysicists have proposed a generalized "neutrino sea." They hold that this flux of uncharged particles comes from the alleged thermonuclear reactions of nearby randomly distributed stars in the universe.

Dudley[10] defines the "neutrino sea" as an energy-rich substrate consisting of:

(a) muon neutrinos e^{o}: rest mass $\simeq$ 0.6 Mev.

(b) electron neutrinos: rest mass $\simeq$ 60 ev.

(c) particle velocity range: a continuum from near zero to near c.
(d) particle density: $\simeq 10^{12}/cm^3$.
(e) energy density estimates: 10^8 to 10^{19} ev/cm^3.

He contends that it is reasonable to conclude that populations of nuclei undergoing what is now called "spontaneous decay" consist of units, each of which is a linear, resonant system. Parametric excitation of these units by interaction with the uncharged particles of the neutrino sea would provide a cause-effect mechanism for the phenomena of radioactivity. The observed "decay constant" is thought to be a complex variable, dependent on the energy content of the atom as a whole and the parameters defining interactions with the neutrino flux. The "decay constant" becomes a stability index rather than a constant. A radioactive atom is thought by Dudley to be analogous to a "set" mouse trap, an energy-rich system in a state of equilibrium, until acted upon by some extraneous mass and/or force.

The disintegration rates of ^{133}Cs clocks have been shown to vary on circumnavigation of the earth in jet planes and Dudley analyzes these changes on the basis of the "neutrino sea" as the primary frame of reference.[11] He indicates that a constant logarithmic decay rate may result from a condition of equilibrium in energy exchange. This equilibrium may be altered by varying the effective velocity of the ^{133}Cs with respect to the neutrino sea, which is in effect streaming past the earth at $\simeq$ 160 km/s.

To illustrate the process Dudley is proposing, let us consider the nuclear reaction called fission. This reaction is considered to take place in a particulate substrate which contributes energy. Potential energy is converted to observable forms through the interaction of high-velocity neutrons with the surrounding energy-rich neutrino flux (ν_e,e^o). I am following Dudley[12] in this example:

Consider ^{235}U to be a linear resonant system subject to parametric excitation by a neutrino flux:

$$^{235}U + \{e^o + \nu_e\} + E_x \rightarrow {}^{231}Th + \propto + E_B$$

($\propto$ decay)

$$^{235}U + \{e^o + \nu_e\} + E_X 2 \rightarrow X + Z + 3n + E_B$$

(spontaneous fission)

In a neutron flux:

$$^{235}U + n \rightarrow {}^{236}U + \nu$$

(n, ν reaction)

$$^{235}U + \{e^{o} + \nu_e\} + n + E_X3 \rightarrow X + Z + 3n + E_B + E_F$$
(fission)

where

E_x = excitation energies from the flux.

X, Z = fission products.

E_B = excess binding energy released.

E_F = energy derived from the flux, the product neutrons acting as catalysts or intermediates in transfer of kinetic energy of $(e^{o} + \nu_e)$ to products and other adjacent nuclei, thus inducing a chain reaction. Net observable Energy:

$E_{net} = E_F + E_B - E_X$

$E_X < E_B$

and $E_B << E_F$

In Dudley's scheme fission reactions are considered to be basically exothermic, but with the major part of the observable energy being supplied by the flux through the product neutrons.

Lee and Yang[13] have predicted that nuclear reactions may be induced by electron neutrinos. Neutrino reactions with stable nuclei have been reported. That of $^{37}Cl + \nu_e \rightarrow {}^{37}Ar + \beta^{-1}$ is an established method of study of high-energy solar (ν_e) neutrinos.

It is postulated that populations of nuclei which are now considered to exhibit spontaneous decay at a constant logarithmic rate, consist of units each of which is a linear resonant system. Parametric excitation of such a unit by an energy input at some critical level or rate may cause the system to react to reduce the effects of this extraneous force, by emitting mass and/or energy, tending toward a more stable system. This is analogous to LeChatelier's Principle: "A system in equilibrium will react to an outside stress such that the effects of the stress will be minimized."

This model, based on the isotropic "neutrino sea" as an energy source, provides a causal basis for the observed exponential $-dN/dt$ in a large population of radionuclides. Uncertainties in predicting the rate of change where N is small would result from the temporary lack of information on the parameters which define the interaction of the medium particularly the number of ν_u and ν_e which constitute the neutrino flux.

B. EVALUATION OF URANIUM-THORIUM-LEAD DATING

Since Boltwood's determination of alleged ages for ten minerals in 1907, the major radiometric time clocks for geological dating used until recent times have been the uranium (U) - thorium (Th) - lead (Pb) series. These series are thought to constitute three independent clocks: U^{238} decays through several elements to give Pb^{206} and 8 helium nuclei; U^{235} gives Pb^{207} and 7 helium nuclei; and Th^{232} gives Pb^{208} and 6 helium nuclei.

1. The Assumptions of the U-Th-Pb "Clocks"

In order for these radioactive decay families to give the age of a mineral (the time elapsed since the mineral crystallized) the following must be known:

a. The half-lives of U and Th must be determined. These quantities can be determined in the laboratory. Henry Faul says: "Most of the common decay constants have assigned errors of 2 percent or less (standard deviation), but the uncertainty in the decay constant of rubidium-87 is much greater. The two values now in use differ by 6 percent."[14]

b. The decay constants of the radioactive minerals must be constant with time. Radiometric dating is predicated on the assumption that throughout the earth's history radioactive decay rates of the various elements (values of the λ's) have remained constant. Is this a warranted assumption? Has every radioactive nuclide proceeded on a rigid course of decay at a constant rate? Very recently this has been challenged by a study involving C^{14} which will be discussed later when considering C^{14} dating.

  (1) At any temperature or pressure, collisions with stray cosmic rays or the emanations of other atoms may cause changes other than those of normal disintegration. It seems very possible that what is called "spontaneous disintegration" of radioactive elements is related in some way to the action of cosmic rays and, if so, the rate of disintegration may vary from century to century according to the intensity of the rays. The evidence for a strongly increasing change in the cosmic ray influx is most favorable in the light of Dr. T. G. Barnes' investigation of the decay of the earth's magnetic field.

  (2) Most geochronologists maintain that pleochroic haloes give evidence that decay constants have not changed. Crystals of biotite, for example, and other minerals in igneous or metamorphic rocks commonly enclose minute specks of miner-

als containing uranium or thorium. The α-particles emitted at high velocity by the disintegrating nuclides interact, because of their charge, with electrons of surrounding atoms which slow them down until they finally come to rest in the host mineral at a distance from their source that depends on their initial kinetic energy and the density and composition of the host. Where they finally stop to produce lattice distortions and defects there generally occurs discoloring or darkening. Each of the 8 α-particles emitted during the disintegration of U^{238} to Pb^{206} produces a dark ring in biotite. Each ring has its own characteristic radius in a given mineral (in this case biotite). This radius measures the kinetic energy, hence the probability of emission of the corresponding α-particle and also the half-life of the parent nuclide accord- to the Geiger-Nuttall law. The Geiger-Nuttall law is an empirical relation between the half-life of the α-emitter and the range in air of the emitted α-particles. If the radii of these haloes from the same nuclide vary, this would imply that the decay rates have varied and would invalidate these series as being actual clocks. Are the radii in the rocks constant in size or are there variable sizes?

Most of the early studies of pleochroic haloes were made by Joly and Henderson. Joly concluded that the decay rates have varied on the basis of his finding a variation of the radii for rocks of the alleged geological ages. This rather damaging result was explained away handily by saying that "enough evidence of correct radii for different geologic periods and sufficient variation in the same period have been obtained that one is forced to look for a different explanation of such variations as were observed by Joly."[15]

Dr. Roy M. Allen made measurements in an excellent collection of samples with haloes. He found that "the extent of the haloes around the inclusions varies over a wide range, even with the same nuclear material in the same matrix, but all sizes fall into definite groups. My measurements are, in microns, 5, 7, 10, 17, 20, 23, 27, and 33."[16]

Most recent studies have been made by Robert V. Gentry. Gentry also finds a variation in the haloes leading him to conclude that the decay constants have not been constant in time.[17]

Incidentally, Gentry points out a very telling argument for an instantaneous fiat creation of the earth. He notes from his studies of haloes: "It thus appears that short half-life nuclides of either polonium, busmuth, or lead were incorporated into halo nuclei at the time of mica crystallization and significantly enough existed without the parent nuclides of the uranium series. For the Po^{218} half-life of ≈ 3 minutes) only a matter of minutes could elapse between the formation of the Po^{218} and subsequent crystallization of the mica; otherwise the Po^{218} would have decayed, and no ring would be visible. The occurrence of these halo types is quite widespread, one or more types having been observed in the micas from Canada (Pre-Cambrian), Sweden, and Japan."[18] The argument seems hard to refute.

So, then, careful scientists have measured variations in halo radii and their measurements indicate a variation in decay rates. The radioactive series then would have no value as time clocks.

(3) As already remarked, among many other requisite assumptions, the validity of radiometric dating procedures depends on the thesis that radioactive decay constants are not altered significantly by any possible environmental influence. Work done by John L. Anderson and George W. Spangler[19] indicates that the generality of the thesis that each nucleus decays independently of all other nuclei within its own species and also of any environment is invalid and that the internuclear effects which apparently cause the deviations of decay from the random expectation appear to be environmentally related.

Experimental work with C^{14}, Co^{60}, Cs^{137} by Anderson and Spangler that provides statistical evidence of non-random emissions suggests that even *mild* changes in the environment may alter the stabilities of the β-emitter of C^{14} and of the β-,r-emitters Co^{60} and Cs^{137}. They have examined half-lives of short-lived isotopes under a series of differing environmental conditions. The evidence suggests markedly different half-lives than those predicted for independent nuclear processors.[20] Anderson and Spangler maintain that their several observations of statistically significant deviations from the (random) expectation strongly suggests that an unreliability factor must be incorporated into age-dating calculations.

c. The final and initial concentrations of U and Th in the sample must be known [see equations (y) and (z)]. The final concentrations of U and Th can be determined quite accurately but the determination of U_o and Th_o is based upon assumptions which do not appear to be valid. It is known that not all lead isotopes and helium in rocks is formed by radioactive decay. Since "original" products are identical to daughter products of decay, there is no really valid way to tell the difference. If theoretical models of nucleosynthesis (formation of elements) and crustal formation are devised and the quantities of Pb^{206}, Pb^{207}, Pb^{208}, and He^4 produced by decay in the sample are known, a set of figures may be calculated regarding the initial concentrations of U and Th in the sample. They are actually only guesses. A guess is made on the basis of a constructed model.

Let us take a look at some of the guessing that is used to find the initial amounts of Pb. Many attempts have been made to find the exact isotopic composition of primordial lead (the lead originally present when the earth was created). The following gives sketches of the three most prominent extrapolations that attempt to find the primordial lead compositions and the age of the earth.

The geochronologist starts from the origin of the earth (about which he knows nothing) and works toward the present. He erects hypothetical systems, makes whatever assumptions he considers warranted, and speculates about what might happen to such models as time passes.

Obviously, the simplest model with which to start would be to assume a solid, homogeneous earth. The composition of the various isotopes of common lead at a time t would be (U^{238}, etc., will be used to denote the number of atoms of that element):

$$\left(\frac{Pb^{206}}{Pb^{204}}\right)_t = \left(\frac{U^{238}}{Pb^{204}}\right)_t (e^{\lambda 238 t} - 1) + \left(\frac{Pb^{206}}{Pb^{204}}\right)_o$$

$$\left(\frac{Pb^{207}}{Pb^{204}}\right)_t = \left(\frac{U^{235}}{Pb^{204}}\right)_t (e^{\lambda 235 t} - 1) + \left(\frac{Pb^{207}}{Pb^{204}}\right)_o$$

$$\left(\frac{Pb^{208}}{Pb^{204}}\right)_t = \left(\frac{Th^{232}}{Pb^{204}}\right)_t (e^{\lambda 232 t} - 1) + \left(\frac{Pb^{208}}{Pb^{204}}\right)_o$$

Pb^{204} is common lead which is not produced by radioactive decay. λ is the respective decay constant, t is the time measured from the origin of the system. The o subscript indicates primordial lead (or the initial amount of that lead). Enough may be said about this model by simply saying that the earth is obviously not homogeneous.

The geochronologist attempts to determine the time of formation of the various lead deposits in the earth by so-called independent geologic means. He then tries to find the development of the isotopic composition of lead as it depends on time. This is still pure guesswork since his independent geologic means are really dependent on his *a priori* assumption of evolution. This assumption guides his dating of geologic events.

A more sohpisticated model—but still as fanciful as the one above—is the Holmes-Houtermans model. They assumed that at the time of the formation the earth differentiated into subsystems each with a characteristic U/Pb ratio. They further assumed that these subsystems remained closed, that is, these subsystems have persisted unmixed to the present time, and that the isotopic composition of the lead in each of them developed from the primordial lead as a function of U/Pb.

Further, let t be the time interval from the origin of the system to the present—the age of the earth. Let t_1 be the time from the moment a lead deposit formed in one of the subsystems to the present. Let the subscripts o and t_1 designate the primordial (or initial) ratios and the ratios at the time of the formation of the deposit, respectively. The isotopic compositions of the lead at the time of formation of the deposit, t_1, are then:

$$\left[\frac{Pb^{206}}{Pb^{204}}\right]_{t_1} - \left[\frac{Pb^{206}}{Pb^{204}}\right]_o = \left[\frac{U^{238}}{Pb^{204}}\right]_{now} (e^{\lambda 238 t} - e^{\lambda 238 t_1}) \qquad (a)$$

$$\left[\frac{Pb^{207}}{Pb^{204}}\right]_{t_1} - \left[\frac{Pb^{207}}{Pb^{204}}\right]_o = \left[\frac{U^{235}}{Pb^{204}}\right]_{now} (e^{\lambda 235 t} - e^{\lambda 235 t_1}) \qquad (b)$$

$$\left[\frac{Pb^{208}}{Pb^{204}}\right]_{t_1} - \left[\frac{Pb^{208}}{Pb^{204}}\right]_o = \left[\frac{Th^{232}}{Pb^{204}}\right]_{now} (e^{\lambda 232 t} - e^{\lambda 232 t_1}) \qquad (c)$$

Divide eq. (a) by eq. (b) and note that $(U^{238}/Pb^{204})_{now} = 137.7(U^{235}/Pb^{204})_{now}$:

$$\frac{\left[\frac{Pb^{207}}{Pb^{204}}\right]_{t_1} - \left[\frac{Pb^{207}}{Pb^{204}}\right]_o}{\left[\frac{Pb^{206}}{Pb^{204}}\right]_{t_1} - \left[\frac{Pb^{206}}{Pb^{204}}\right]_o} = \frac{1}{137.7}\left[\frac{e^{\lambda 235 t} - e^{\lambda 235 t_1}}{e^{\lambda 238 t} - e^{\lambda 238 t_1}}\right]$$

Write the right hand expression as $Q = 137.7 \left[\frac{e^{\lambda 238 t} - e^{\lambda 238 t_1}}{e^{\lambda 235 t} - e^{\lambda 235 t_1}}\right]$

There are three unknowns in the above equation: t, $(Pb^{207}\backslash Pb^{204})_o$, and $(Pb^{206}/Pb^{204})_o$ since t_1 may be specified arbitrarily. This still is an impossible situation for a solution for t is dependent on the assumptions regarding the primordial lead ratios.

To see the thinking involved here, note the thinking of Henry Faul as he writes: "If one assumes that the solar system condensed from a primordial cloud, it follows that the materials of planets, asteroids, and meteorites have a common origin. Iron meteorites contain some lead but only infinitesimal traces of uranium and thorium, and therefore the lead is uncontaminated by radiogenic lead and can be regarded as a good sample of primordial lead. Table 6-1 lists the isotopic composition of lead extracted from some iron meteorites. These data now can be used as $(Pb^{207}/Pb^{204})_o$ and $(Pb^{206}/Pb^{204})_o$ in the Houtermans equation, and all that remains to be found to permit a calculation of the age of the earth is a lead sample from a closed, subsystem of well-known age."[21]

Again, there exists a conglomeration of "ifs" and suppositions in all this. One may note: (A) who knows how the solar system was formed? (B) The composition and structure of the planets themselves are radically different. (C) Certainly, asteroids are very different from planets. (D) As meteoroids travel through space, much of the Pb in the iron is produced not by decay of U and Th, but by cosmic ray incidence. Thus, the determination of $(Pb^{207}/Pb^{204})_o$ and $(Pb^{206}/Pb^{204})_o$ by determination of Pb^{207}/Pb^{204} and Pb^{206}/Pb^{204} in meteorites seems most unreliable.

Alpher and Herman set up a model where the lead source in the crust is completely mixed. In this model

$$Q = 137.8\ \frac{(e^{\lambda 238t} - 1)}{(e^{\lambda 235t} - 1)}$$

Each school tries to avoid conflict by applying its model only to a particular type of anomaly. However, aside from the difficulty of finding initial quantities by using models that are pure guesses about the early crust of the earth, there are also basic difficulties in finding the amounts of Pb^{206}, Pb^{207}, Pb^{208}, and He^4 produced by radioactive decay for the following reasons:

(1) It is possible and very likely that not all of the Pb^{206}, Pb^{207}, Pb^{208}, and He^4, the decay products of U and Th, found in the samples were produced by radioactive decay. A part of each of the lead isotopes may have been original lead. The helium could have been present when the rocks were formed and not produced at all by radioactive disintegration.

(2) There are physical and chemical changes taking place in the crust of the earth that have nothing to do with radioactive decay. For example, helium escapes at a very rapid rate from the rocks into the atmosphere of the earth. It has been

estimated that 10,000 to 100,000 tons of helium are exuded into the atmosphere from the rocks each year. U is being carried into the oceans from the rocks at a rate variously estimated from 10,000 to 5,000,000 tons per year.[22] These depletions take place near the surface which is where the sampling takes place. Also, meteors, meteorites, and micrometeorites are bringing U and Th into the atmosphere and onto the surface of the earth. Further, U and Th are brought to the surface by volcanic action. These changes will affect the correctness of the measured amounts of U and Th, thus producing error in the equations for the time calculations. If the amount of U or Th is smaller because of these physical chemical effects, this will give appearance of greater age by radioactive decay.

Obviously, there are major uncertainties and likely errors which can lead to vastly incorrect ages.

d. Discordant U-Pb ages are often obtained for chondrites. The work of N.H. Gale and J.W. Arden showed that four important chondrites contain radiogenic Pb whose isotopic composition according to the conventional interpretation is not supported by their uranium content. They suggested that a wrong choice had been made of the initial Pb isotopic composition from which the meteorite Pb's have come. M.C.B. Abranches[23] along with Arden and Gale have initiated work on Richardton and Farmington chondrites to try to determine the initial Pb isotopic compositions in these meteorites, to show the reality of the apparent excess radiogenic Pb seen in many meteorites when the Cañon Diablo triolite Pb is used as the initial Pb isotopic composition and to establish that the apparent excess radiogenic lead is real and not the result of contamination of the meteorites with terrestrial lead.

Many investigators believe that the excess lead found in these studies is due to terrestrial contamination because of imperfect sampling or terrestrial contamination prior to analysis. However, Abranches, Arden, and Gale[24] maintain that "to attribute to contamination all cases where the U-Pb system in whose meteorite samples appear to be discordant is merely an assumption which may hide a real phenomenon in need of explanation." They maintain that the evidence is largely against the terrestrial contamination of most of the samples of the Richardton and Farmington meteorites they used. The problem of variable apparent initial Pb isotope composition seems real and not explainable by contamination.

Some of the conclusions reached by Abranches, Arden, and Gale[25] are as follows:

(1) All samples show an apparent excess radiogenic Pb when Cañon Diablo triolite is used as an initial.

(2) The calculated apparent initial Pb isotope compositions show wide variations in 1-to 2-g samples of a given meteorite.

(3) Evidence is presented to show that the excess Pb cannot, in the majority of instances, be explained by terrestrial contamination.

(4) Both the single- and two-stage evolution models are not applicable as there is no unique initial Pb isotopic composition, and the data are inconsistent with the ordinary two-stage model and both the linear three-stage models for each meteorite.

(5) The apparent excess Pb and apparent variable initial Pb isotopic compositions appear to be real.

e. Appreciable differentiation no doubt has occurred. Many of the uranium salts are water soluble, especially in water containing dissolved oxygen under pressure. These salts migrate with water very readily, both on the surface and underground, until they enter a reducing environment. Hurley has pointed out that the radioactive components of granites reside almost entirely on the grain surface and may readily be leached from the granite. It would appear that a tremendous amount of differentiation has taken place during the lifetimes of the rocks.

2. Criticism of the U-Th-Pb Methods

Laying aside for the moment the apparent unsolvable problems entailed by the assumptions, let us consider the methods themselves. Do these systems really tell time? Do they possess a time or age index, or have the rocks been in existence just long enough to get them started and no more? If most of the lead were present initially, then only a small amount of the lead would be radiogenic. This would indicate that the "clocks" have just started "ticking." There is no physical or chemical difference in original and radiogenic Pb. Let's see what the data show.

There are two classes of U-Th-Pb "clocks" that are ordinarily used as clocks: (1) the lead-rich minerals containing very small quantities of U and Th, but a large quantity of Pb. It is assumed that the lead-rich minerals (called common leads) were "fed" by U and

Th decay before the mineralization when the rock sample was formed. Thus, these minerals had an initial lead isotope composition, and the lead composition in the rock has changed with time from that initial composition. A method for checking on the time index is discussed below which eliminates the necessity of knowing precisely the initial amount of Pb in the minerals. (2) The U-Th-rich minerals containing very little lead, but much U or Th and called radioactive minerals.

a. **Lead-Rich Minerals**

Dr. Melvin A. Cook has come up with a rather ingenious approach to determine whether or not lead-rich minerals are intrinsically radiometric clocks.[26] He utilizes a mathematical expression which contains a time-index or age-indicator (a mathematical quantity which should change as time increases). He considers the ratio of the change of Pb^{206} with time to the change of Pb^{207} with time, or,

$$Q = \frac{y - y_o}{z - z_o}$$

where y, y_o are the present and initial (or primordial ratios) of Pb^{206} to Pb^{204}, respectively, and z, z_o are the present and initial (or primordial) ratios of Pb^{207} to Pb^{204}. Instead of using the initial values of these ratios at time $t = 0$, he uses the ratios at some arbitrary time t_o later.

Dr. Cook made a plot of the present-day ratios of $Pb^{206}/Pb^{207} = x$ versus $Pb^{206}/Pb^{204} = y$. To find y_o he took the smallest value of y from this curve. Now, z_o may be found from $z_o = y_o/x_o$ since $(Pb^{207}/Pb^{204})_o = (Pb^{206}/Pb^{204})_o/(Pb^{206}/Pb^{207})_o$. It does not really matter that that part of y_o and z_o may be contributed by radiogenic lead produced by U decay before these isotope ratios were reached, because all larger values of y and z would, according to the conventional theory, have passed through y_o and z_o at some time t_o anyhow.

Thus, $Q = \dfrac{y - y_o}{z - z_o}$ can be calculated for samples taken from the earth, since y and z can be measured in the laboratory and y_o and z_o have been found as described above. Thus Q has a time-index or age-indicator in it since

$$Q = \frac{\Delta y}{\Delta z} = \frac{U_0^{238} - U_1^{238}}{U_0^{235} - U_1^{235}} = \frac{U^{238}}{U^{235}} \left[\frac{U_0^{238}/U^{238} - U_1^{238}/U^{238}}{U_0^{235}/U^{235} - U_1^{235}/U^{235}} \right]$$

$$= 137.8 \left[\frac{e^{\lambda 238 t_0} - e^{\lambda 238 t_1}}{e^{\lambda 235 t_0} - e^{\lambda 235 t_1}} \right]$$

where U is the number of uranium atoms now, U_o the number at time t_o ago (when $y = y_o$ and $z = z_o$) and U_1 the number at time t_1 (the time of mineralization). Since Δy and Δz are the amounts of Pb^{206} and Pb^{207}, respectively, due to radioactive decay only, these quantities are just equal to the number of atoms of U^{238} and U^{235} converted by radioactive decay from their primordial amounts to Pb^{206} and Pb^{207}, respectively. Thus, Q has in it a time-index or age-indicator for

$$Q = 137.8 \frac{e^{\lambda 238 t_0} - e^{\lambda 238 t_1}}{e^{\lambda 235 t_0} - e^{\lambda 235 t_1}} \quad \text{since } \lambda_{238} \neq \lambda_{235}.$$

Q should change as t_1 changes since the times of mineralizations of the different rocks are radically different by the conventional evolutionary scheme. The time index should reflect the alleged age of the various rocks.

When Dr. Cook (see his classic study, *Prehistory and Earth Models*) analyzed the various collections of data by different investigators, all holding the evolutionist view regarding the age of the earth, he found that there was no detectable trend at all for lead-rich minerals of one age to differ from those of another age. The time-index that was built into the mathematical expression Q should have changed. According to the Holmes-Houtermans model, Q should vary from 2.5 in lead-rich rocks allegedly 2.5 billion years old to 5.9 in modern lead-rich rocks. According to the Alpher-Herman model, it should have varied from 5.9 for rocks allegedly 2.5 billion years old to 21.5 in a modern rock.[27] Dr. Cook used data from the analysis of 500 samples of widely-differing alleged ages and found that there was no definite trend in average value of Q. Thus, for Q there is no systematic variation whatever from one kind of formation or geologic "period" to another or from one locality to another. This so-called age index has all different values even for samples that are of the same supposed geologic period.

This means that the lead isotope ratio method for dating samples containing lead-rich minerals can give no ages and is useless as a time clock. The above results could alternatively be interpreted as meaning that the ages of all the rocks containing lead-rich minerals are too small to have produced appreciable changes in Q.

b. **Radioactive Minerals**

A close look at the radioactive minerals reveals a major and fatal flaw in the alleged U-Th-Pb time clocks. Ages (t's) are found using the following ratios:

t_1 from Pb^{206}/Pb^{207}, t_2 from Pb^{207}/U^{235}, t_3 from Pb^{206}/U^{238}, t_4 from Pb^{208}/Th^{232}.

The ratio Pb^{206}/Pb^{207} may be used to tell time, since the amount of Pb^{207} produced in a mineral per year drops off much more rapidly than the amount of Pb^{206} produced per year. The equation for the time by the lead-lead ratio is

$$\frac{Pb^{207}}{Pb^{206}} = \frac{U^{235}}{U^{238}} \frac{(e^{\lambda 235 t_1}-1)}{(e^{\lambda 238 t_1}-1)} .$$

From the various age determinations, it turns out that $t_1 > t_2 > t_3 > t_4$; or, if the Pb^{208}/Th^{232} is taken as the standard clock, Pb^{206}/U^{238} gives a 12% longer age, Pb^{207}/U^{235} gives an 18% longer age, Pb^{206}/Pb^{207} gives a 35% greater age. A crucial question is: Why should these "clocks" give substantially different ages?

It is believed that chemical differentiation could explain to to some extent the difference between the Pb^{208}/Th^{232} "clock" and the other clocks. Uranium is more easily leached out of different radioactive minerals than thorium is, but why should the other clocks be radically different?

Dr. Cook has suggested that the explanation for these ages being very different is that Pb^{206} is being converted to Pb^{207} and Pb^{208} by neutron reactions.[28] There is no tendency for separation of the natural isotopes in the crust of the earth. It would seem that the only other possibility of explaining why these clocks give different ages would be nuclear transformations of one lead isotope into another by reactions of the lead nuclei with neutrons. This process is illustrated in Fig. 4.

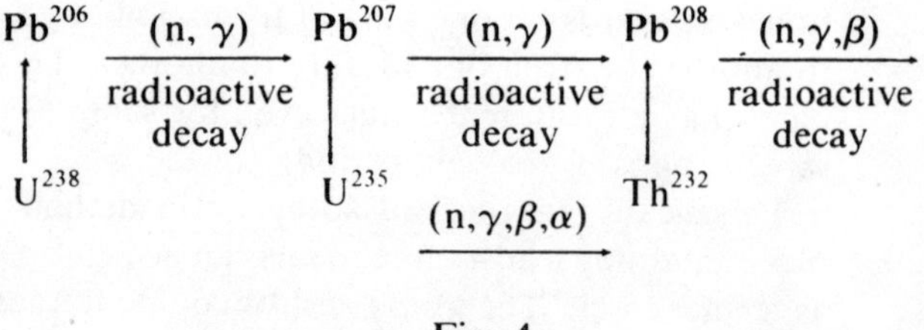

$$\begin{array}{ccccccc} Pb^{206} & \xrightarrow[\text{radioactive decay}]{(n,\gamma)} & Pb^{207} & \xrightarrow[\text{radioactive decay}]{(n,\gamma)} & Pb^{208} & \xrightarrow[\text{radioactive decay}]{(n,\gamma,\beta)} & Bi^{209} \\ \uparrow & & \uparrow & & \uparrow & & \\ U^{238} & & U^{235} & \xrightarrow{(n,\gamma,\beta,\alpha)} & Th^{232} & & \end{array}$$

Fig. 4

The horizontal arrows noted with (n,γ) above them indicate neutron-promoted reactions. There are three reasons cited by Dr. Cook for thinking this is the explanation.

(1) The age determined by the ratio Pb^{206}/U^{238} is ordinarily less than that by Pb^{207}/U^{235} when neutron reactions are involved because neutron bombardment converts more Pb^{206} to Pb^{207} than Pb^{207} to Pb^{208}. The effect is very pronounced in the Pb^{206}/Pb^{207} ratio, giving the largest "age" by this "clock" than the others. The larger the Pb/U ratio and the smaller the Pb^{206}/Pb^{207} ratio, the greater the "age" of the sample.

Dr. Cook has referred to two well-documented cases to back up this contention. Quoting his words regarding these cases: "Consider, for example, the uranium ore body of Shinokolobwe, Katanga. In this ore quite generally Pb^{204} is absent, as also is Th^{232}. Yet there is present in this ore some Pb^{208}! From where does it come? The absence of Pb^{204} (the nonradiogenic lead) implies that there is no original lead in this ore; apparently, all of the lead is radiogenic.

If there were any original lead at all the prominent constituent would be Pb^{204}. On the other hand, since there is no Th^{232} either, the Pb^{208} could not have come from Th^{232} decay. Therefore, it probably came from the (n,γ) reaction: $Pb^{207} + n \longrightarrow Pb^{208} + \gamma$."[29] He then proceeds to put a correction factor into the age calculation to account for the (n,γ) reactions producing the Pb^{208}. This reduces the alleged age by conventional theory of this sample ore from 600 million years to nearly zero. This was also done for Martin Lake ore taken from the Canadian Shield. In this case, the correction factor for neutron reactions reduced the alleged age from 1.7 billion years to nearly zero.

The method of correction for these two "ages", following Dr. Cook's approach to the problem, is as follows:

(a) **Shinokolobwe, Katanga:**

1—Observed ratio $Pb^{206}/Pb^{207} = 94.2/5.72 = 16.5$.

2—Pb^{208} averaged 0.08% of the total lead, but if 0.08% from the Pb^{207} (no other source), 16.5 times more Pb^{207} or 1.3% (0.08 × 16.5) should have been produced by the reaction $Pb^{206} + n \longrightarrow Pb^{207} + \gamma$ since there was 16.5 times more Pb^{206} than Pb^{207}.

3—The corrected ratio of $Pb^{206}/Pb^{207} = \dfrac{(94.2 + 1.3)}{(5.72 + 0.08 - 1.3)} = 21.1$.

4—This corrected ratio says the corrected age should be practically zero since $Pb^{206}/Pb^{207} = 21.5$ for modern radiogenic lead.

(b) **Martin Lake:**

1—Observed ratio $Pb^{206}/Pb^{207} = 90.4/9.1 = 9.9$.

2—An average of 0.53% Pb^{208} but enough Th to generate less than 1% of this Pb^{208} or then an average of 0.52% Pb^{208}. So if 0.52% of the Pb^{208} came from Pb^{207}, 9.9 times more Pb^{207} or 5.2% (0.52 × 9.9) should have been produced by the reaction $Pb^{206} + n \longrightarrow Pb^{207} + \gamma$ since there was 9.1 times more Pb^{206} than Pb^{207}.

3—The corrected ratio of $Pb^{206}/Pb^{207} = \frac{(90.4 + 5.2)}{(9.1 + 0.52 - 5.2)} = 20.7$.

4—This corrected ratio comes back down to modern lead.

Almost all large uranium mines have been found to have the anomalous conditions found in the above situations.

Apparently then, the (n,γ) reactions change the Pb^{206}/Pb^{207} ratio rather radically, causing it to give a large apparent age. The indication from these ores is that the rocks haven't really been around long enough to cause appreciable decay products. Instead of radioactive decay producing the major part of the "radiogenic" Pb, neutron reactions can account for it.

(2) As a further test of the idea that neutron reactions have produced changes in the isotopic ratios causing the long ages of conventional theory, Dr. Çook set up another mathematical expression.[30] This equation he called the P-ratio,

$$P = \frac{\Delta w}{\Delta y}\left[\frac{U^{238}}{Th^{232}}\right]\left[\frac{e^{\lambda 238 t}-1}{e^{\lambda 232 t}-1}\right]$$

where $w = \frac{Pb^{208}}{Pb^{204}}$, $y = \frac{Pb^{206}}{Pb^{204}}$, $\Delta w = w - c$, and $\Delta y = y - a$.

a, c are the values of y, w at some time measured backward from the present. This P-ratio should be 1 if conventional theory is correct and the main part of the lead is produced by radioactive decay; but this ratio would vary from infinity (as Th/U goes to zero) to zero (as Th/U goes to infinity), according to Dr. Cook's neutron-reaction theory. That this P-ratio should equal 1 by conventional theory may readily be seen as follows:

$$P = \frac{\Delta w}{\Delta y}\left[\frac{U^{238}}{Th^{232}}\right]\left[\frac{e^{\lambda 238t}-1}{e^{\lambda 232t}-1}\right]$$

$$= \frac{\left[\frac{Pb^{208}}{Pb^{204}}\right]_{now} - \left[\frac{Pb^{208}}{Pb^{204}}\right]_{o}}{\left[\frac{Pb^{206}}{Pb^{204}}\right]_{now} - \left[\frac{Pb^{206}}{Pb^{204}}\right]_{o}} \frac{U^{238}(e^{\lambda 238t}-1)}{Th^{232}(e^{\lambda 232t}-1)}$$

$$= 1$$

since $\left[\frac{Pb^{206}}{Pb^{204}}\right]_{now} - \left[\frac{Pb^{206}}{Pb^{204}}\right]_{o} = U^{238}(e^{\lambda 238t}-1)$

and $\left[\frac{Pb^{208}}{Pb^{204}}\right]_{now} - \left[\frac{Pb^{208}}{Pb^{204}}\right]_{o} = Th^{232}(e^{\lambda 232t}-1)$

If the neutron reactions play a role in the production of Pb, it would be expected that the P-ratio would vary from zero to infinity.

When the P-ratios were calculated for many samples, Dr. Cook found that P varied from near zero to infinity.[31] Therefore, it seems eminently reasonable to conclude that U-Th-Pb systems either are not time clocks or are telling a time too short to be measureable.

(3) In air, the ratio N^{14}/N^{15} is 269. However, the ratio of these nitrogen isotopes in radioactive minerals is 162. This means that the N^{15} content in these minerals is higher than air. This substantial change of ratio seems to be caused by neutron interactions, the neutrons having come from the radioactive ore body. It seems certain that if neutron reactions can produce such a large change in the N^{14}/N^{15} ratio, it could surely change the Pb^{206}/Pb^{207} ratio by a large amount also.

Dr. Cook states that "in a massive ore body the level of the neutron flux might be only a millionth (or less) as great as it is in an experimental fast neutron pile. Yet even at this low level of activity, the neutron flux would be ample to upset completely the uranium-thorium-lead time clocks and generate 'billion-year-old' minerals in a few thousand years!"[32]

(4) It has been reported[33] that an abundance of xenon-131 found in lunar samples can be explained by the transmutation of barium-130 to copious quantities of xenon-131 un-

der neutron bombardment. The transmutation of barium-130 was observed at Lawrence Livermore Laboratory's 100 m.p.v. electron-positron accelerator, which can produce a spectrum of neutrons duplicating that believed to exist at the lunar surface.

C. CRITICISM OF THE POTASSIUM-ARGON METHOD

The potassium-argon clock has been heralded loudly and was supposed to have great prospects. But, as of late, it has fallen into some bad times. There are some very serious objections to using the potassium-argon decay family as a radiometric clock. This is, of course, very harmful to the position of those holding the theory of sea-floor spreading since their time scale has been calculated using mainly K^{40}/Ar^{40} dates. About 11% of K^{40} decays by electron capture and gamma ray emission to Ar^{40} and the remaining 89% of the K^{40} decays by β-particle emission to form Ca^{40} (see Fig. 3-b). The geochronologist considers the Ca^{40} of little practical use in radiometric dating since common calcium is such an abundant element and the radiogenic Ca^{40} has the same atomic mass as common calcium. Potassium is present in most geological materials, making this system highly useful if it really works.

For this system to work as a clock, the following four criteria must be fulfilled:

1. The decay constant and the abundance of K^{40} must be known accurately.
2. There must have been no incorporation of Ar^{40} into the mineral at the time of crystallization or a leak of Ar^{40} from the mineral following crystallization.
3. The system must have remained closed for both K^{40} and Ar^{40} since the time of crystallization.
4. The relationship between the data obtained and a specific event must be known.

In areas where tremendous tectonic activity has taken place, highly discordant values for the ages are obtained. The difficulties associated with these criteria are numerous and may be listed as follows:

1. There seems to be a great deal of question regarding the branching ratio for K^{40} into Ar^{40} and Ca^{40}. The value that has been used for Ar^{40}/Ca^{40} has varied from 0.12 to 0.08. But, as Dr. Melvin Cook[34] points out, the value is not really known. The observed value is between 0.11 and 0.126, but in order to match K-Ar ages, which average somewhat higher than the U-Th-Pb ages, to the latter ages, the value 0.08 is arbitrarily taken. However, this does not remedy the situation and the ages are still too high. The geochronologists credit this to "argon leakage".

2. There is far too much Ar^{40} in the earth for more than a small fraction of it to have been formed by radioactive decay of K^{40}. This is true even if the earth were really 4.5 billion years old. In the atmosphere of the earth, Ar^{40} constitutes 99.6% of the total argon. This is around 100 times the amount that would be generated by radioactive decay over the hypothetical 4.5 billion years.[35] Certainly this is not produced by an influx from outer space. Thus, it would seem that a large amount of Ar^{40} was present in the beginning. Since geochronologists assume that errors due to presence of initial Ar^{40} are small, their results are highly questionable.

3. Argon diffuses from mineral to mineral with great ease. It leaks out of rocks very readily and, therefore, can move from down deep in the earth, where the pressure is large, and accumulate in an abnormally large amount in the surface where our rock samples for dating are found. They would all have excess argon due to this movement. This makes them appear older. Rocks from deeper in the crust would show this to a lesser degree. Also, since some rocks hold the Ar^{40} stronger than others, some rocks will have a large apparent age, others smaller ages, though they may actually be the same age. If you were to measure Ar^{40} concentration as function of depth, you would no doubt find more of it near the surface than at deeper points because it migrates more easily from deep in the earth than it does from the earth into the atmosphere. It is very easy to see how the huge ages are being obtained by the K^{40}-Ar^{40} radiometric clock, since surface and near-surface samples will contain argon due to this diffusion effect.

4. Many of the rocks seem to have inherited Ar^{40} from the magma from which the rocks were derived. Volcanic rocks erupted into the ocean definitely inherit Ar^{40} and helium, and thus when these are dated by the K^{40}-Ar^{40} clock, old ages are obtained for very recent flows. For example, lavas taken from the ocean bottom off the island of Hawaii on a submarine extension of the east rift zone of Kilauea volcano gave an age of 22 million years, but the actual flow happened less than 200 years ago.[36]

   Some geochronologists believe that a possible cause of excess argon is that argon diffuses into the mineral progressively with time. Significant quantities of argon may be introduced into a mineral even at pressures as low as one bar.

   If such "wild" ages as mentioned above are obtained for pillow lavas, how "wild" are those from deep-sea drilling out in the Atlantic where sea-floor spreading is supposed to be occurring?

5. Potassium is found to be very mobile under leaching conditions. As much as 80% of the potassium in a small sample of an iron meteorite was removed by running distilled water over it for 4 1/2 hours. This could move the "ages" to tremendously high values. Ground-water and erosional water movements could produce this effect naturally.
6. Rocks in areas having a complex geological history have many large discordances. In a single rock there may be mutually contaminating, potassium-bearing minerals.
7. There seems to be some difficulty in determining the decay constants for the K^{40}-Ar^{40} system. Geochronologists use the branching ratio as a semi-empirical, adjustable constant which they manipulate instead of using an accurate half-life for K^{40}.

D. CRITICISM OF THE RUBIDIUM-STRONTIUM DATING METHOD

The geochemical properties of rubidium (Rb) and strontium (Sr) are such that any mineral containing Rb^{87} is very likely to contain amounts of inherited Sr^{86} and Sr^{87} as well as radiogenic Sr^{87} produced by radioactive decay of Rb^{87}. The Rb^{87} isotope decays by β-emission to Sr^{87}. The decay constant is uncertain, but the value commonly used is $\lambda = 1.39 \times 10^{-11}/yr$, making the half-life 49.9 billion years. This Rb^{87}-Sr^{87} clock is used particularly in attempts to date metamorphic rocks. However, its ability as a time clock is just as questionable as the other methods and, considering the unrealistic assumptions made, seems not to be a clock at all. Let's take a look at some of the assumptions and difficulties involved.

1. It seems utterly impossible to determine the initial concentration of Sr^{87} atoms. The Sr^{87} present in the earth's crust is at least ten times more abundant than it should be if it were all formed by Rb^{87} decay over 5 billion years.[37] This means that most Sr^{87} is nonradiogenic. There is really no valid way of determining what the initial amounts of Sr^{87} in rocks were. There is much juggling of numbers and equations to get results in agreement with the U-Th-Pb "clocks". In all these radioactive clocks, all methods are made to give values that fit the evolutionist's belief as to the age of the earth and the ages of the various geological events. The reason that the various dating methods give similar ages after "analysis" is that they are made to do so. In the case of the initial Sr^{87}/Sr^{86} ratios, these values can be adjusted so that any age desired is obtainable.
2. In the Rb^{87}-Sr^{87} system, the ease of diffusion of Sr^{87} is completely ignored. If these elements can move about, then the whole dating

technique has no value at all, since there is no way of deciding on the amounts of Rb and Sr involved in the diffusion process. Many geochronologists believe that in numerous cases the whole-rock may be closed to diffusion of Rb and Sr, but that the individual minerals making up the whole-rock are open to diffusion. This does not seem even remotely reasonable. Of course, the reason for this device was to get out of the difficulty of very discordant ages between whole-rock analyses and mineral analyses. When individual minerals in a rock were used, various ages were obtained. These mineral age values were considered discordant since they differed radically from a value of the age using the whole-rock sample instead of certain constituent minerals. This is what is expected on the basis of ready diffusion of Sr^{87} in a rock and between rocks.

3. The whole-rock Rb-Sr ages which are considered of the greatest importance in geochronology have many serious faults. After explaining how this whole-rock age is determined, we will take a look at the faults.

For a rock containing both initial and radiogenic strontium, the total amount of Sr^{87} is given by:

$$(Sr^{87})_{now} = (Sr^{87})_{O} + (Sr^{87})\text{ radiogenic}$$
$$= (Sr^{87})_{O} + (Rb^{87})_{now}\,(e^{\lambda t} - 1).$$

Referring the Sr^{87} and Rb^{87} to Sr^{86} since what is really measured is a ratio of elements, we get:

$$\left[\frac{Sr^{87}}{Sr^{86}}\right]_{now} = \left[\frac{Sr^{87}}{Sr^{86}}\right]_{O} + \left[\frac{Rb^{87}}{Sr^{86}}\right]_{now}(e^{\lambda t} - 1)\,.$$

The subscript o refers to the primordial amounts. This may be written as $y = mx + b$ where $y = (Sr^{87}/Sr^{86})_{now}$, $X = (Rb^{87}/Sr^{86})_{now}$, $m = e^{\lambda t} - 1$, $b = (Sr^{87}/Sr^{86})_{O}$. This is the equation of a straight line, called in this case an isochron. When plotted for different samples from the same rock, the slope of the isochron supposedly gives the age of the rock and the intercept on the Sr^{87}/Sr^{86} axis supposedly gives the initial amount of Sr^{87}. Well, it looks good, but the method has clay feet. Here are the clay feet.

a. Frequently there is too much Sr^{87} in the whole rock to have come about radiogenically from Rb in several billion years, so that radiogenic Sr^{87} represents only a small percentage of the total Sr^{87}, and the necessity of subtracting away most of the Sr^{87} may introduce a large error in measurement. In other words, practically all the sample may be nonradiogenic Sr^{87}.
b. For the isochron technique mentioned above to be valid, (a) all samples must have the same initial Sr^{87}/Sr^{86} ratio; (b) they must

all have the same age; and (c) the rock must have acted as a closed system. Regarding the criterion that the rock has been a closed system: The diffusion of Rb and Sr and contamination of the samples involving these elements can occur quite readily; thus it is fairly certain that none of the rocks are closed systems. Certainly, if contamination and diffusion have taken place, how is one to decide the ages?

Now, concerning the assumption that the samples had the same initial Sr^{87}/Sr^{86} ratio, some pertinent remarks may be made. First, if it is assumed that there is a uniform distribution of Sr^{87} in the rock, then it is assumed that there is also a uniform distribution of Rb^{87}. But, of course, this is not assumed by the geochronologist since there would, by conventional theory, have to be a clustering of his points at one position on a Sr^{87}/Sr^{86} vs. Rb^{87}/Sr^{86} graph. Because of this, there would be no straight line or isochron defined at all, and no way to find the initial Sr^{87}/Sr^{86} ratio. Further, if contamination has occurred, then both the Sr^{87}/Sr^{86} ratio and the Rb^{87}/Sr^{86} ratio will cause the isochron equation to give an entirely erroneous value for (Sr^{87}/Sr^{86}) initial which in turn leads to an incorrect age. Dr. Cook has pointed out that the obtaining of the isochrons is better explained as a natural isotopic variation effect, since similar curves are obtained for plots of Fe^{54}/Sr^{86} vs. Fe^{58}/Sr^{86} which are known not to be time functions[38] since these ratios have nothing to do with radioactivity because these isotopes are not radioactive. There is no way to correct for this natural isotopic variation since there is no way to determine it. This renders the Rb^{87}-Sr^{87} series useless as a clock.

4. Sr^{87} disappears in many cases by ion exchange processes. Many statements are made in the writings of geochronologists that an age is too high or too low. Whether an "age" is too high or too low depends on what has been assumed to be the correct age. The U-Th-Pb series are assumed to give the right values and are used as a standard. In radiometric dating, the evolutionist works in a circle since everything he does depends on a set of assumptions where validity cannot be known.
5. Although the mantle is very deep within the earth it is of great importance in affecting the surface of the earth. The mantle is believed by many to yield the magma to form the crust beneath the ocean and, perhaps, to add new rock to the continental crust. The basalts of the crust are believed to have been derived from the mantle.

   Geochemists, relying heavily on isotopic analyses of the basalts, maintain that the mantle is not a "single, chemically uniform layer of rock,"[39] but is chemically heterogeneous. In this view it is held that the isotopes in a rock depend directly on their proportions in the mantle rock from which the basalt was believed to have been derived by melting. The melting process has no effect on the isotopic composition of a basalt but the same part of the mantle can produce rock with different chemical characteristics depending on how and to what extent the mantle is melted.

Investigators of the French-American Mid-Ocean Study found lava flows lying close by each other on the Mid-Atlantic Ridge at 36° N latitude whose chemical compositions differ markedly. It would appear that there may be separate sources for this basaltic rock.

Presently the island rocks along the Mid-Atlantic Ridge have been studied most thoroughly in this regard. The ratio of Sr-87 to Sr-86 (Sr^{87}/Sr^{86}) is generally higher on Iceland and the Azore Islands than it is on the Mid-Atlantic Ridge.[40] This is believed to be due to chemical differences of separate parts of the mantle.

The isotopic systems of lead/uranium and neodymium/samarium are believed to also indicate chemical differences in the mantle. Extensive sampling (dregding, collection from research submarines, deep-sea drilling) shows considerable isotopic variation along the Mid-Atlantic Ridge. Much of the magma believed extruded along the Mid-Atlantic Ridge shows much variation from what is considered "normal."

Many investigators believe there may be multiple sources for basalts in the mantle with widely different chemical composition. Obviously, this casts tremendous doubt on "isotopic dates." The calculated lead isotope "age" will depend on how the samples used in the calculation are selected. The mantle apparently has been changing with time. Assumptions galore are required.

6. The world's volcanic rocks are generally classed as oceanic or continental. The tectonic environment leading to volcanism seems to differ between and within these two classes. Geologists have tries to correlate the mode of formation with a distinctive geochemical imprint but have not been particularly successful. However, it appears that continental volcanics ordinarily possess higher and more variable Sr^{87}/Sr^{86} ratios than oceanic volcanics.[41]

   It is generally held that the Sr isotope values for oceanic volcanic rocks show the isotopic compositions of the mantle. The mantle-derived magmas in the continents have crossed 30 to 40 km of radiogenic sialic crust and the high and variable Sr isotopic ratios are usually ascribed to crustal contamination. Further it is thought that the higher Sr ratios may have been inherited directly from subcontinental mantle possessing anomalous Sr isotopic compositions. This situation leads to all sorts of confusion in determining initial Sr^{87} in the rocks. It is maintained that the isotopic compositions "correlate with Rb/Sr ratios to form pseudoisochrons which give ages grossly in excess of the true age of volcanism". C. Brooks, D.E. James, and S.R. Hart claim that the high and variable Sr^{87}/Sr^{86} ratios are "the direct result of (i) large-scale 'primary' heterogeneities frozen into the ancient lithosphere and (ii) small-scale 'secondary' heterogeneities resulting from disequilibrium of radiogenic daughter products."[42]

Brooks, James and Hart try to use what they have called the pseudiosochrons to determine mantle processes which they believe have caused the pseudoisochrons for Rb/Sr dating. They proceed to "correct" Sr^{87}/Sr^{86} for what they propose are the true ages of the rocks. Their use of correlation theory and regression analysis indicate that "most of these pseudoisochrons have slopes significantly different from zero at confidence levels up to 95 percent (in some cases up to 99.9 percent) and that they define excess "ages" ranging from 70 million years to more than 3 billion years."[43] The so-called isochrons appear then to be quite meaningless in obtaining true ages in accord with the conventional theory. The data (even when interpreted from the evolutionist view) indicates inherited ages which cannot be removed without a knowledge of crystallization ages.

E. CRITICISM OF THE RADIOCARBON CLOCK

1. Physics of the C^{14} Dating Method

The C^{14} dating method has a great reputation for accuracy, especially for the dating of archeological artificts and related geological occurrences. This method has won great fame for its inventor, William Libby won the Nobel Prize for his work along this line. Let us see how the C^{14} dating method works. When cosmic radiation impinges upon the earth's upper atmosphere, neutrons are produced as a result of the collisions of these high-speed protons (the cosmic rays) with particles in the upper atmosphere. The neutrons that are produced as a result of these collisions move with very high velocities and collide with N^{14} atoms in the atmosphere, particularly in the stratosphere. Upon the collision with the N^{14} particles, C^{14} is produced. This C^{14} is an unstable form of carbon, the radioactive form. The most abundant non-radioactive form of carbon is C^{12}. Radio-

active carbon formed in the atmosphere decays and forms N^{14} [according to the equation $(C^{14}/C^{12})now = (C^{14}/C^{12})_{O}e^{-\lambda_{14}t}$].The exchange of carbon-14 between the atmosphere and the organism takes place continually while an organism, such as a tree or a human being, is alive. When the organism dies, this exchange no longer occurs. The C^{14} in the organism continues to decay radioactively and the ratio of C^{14} to C^{12} becomes smaller and smaller as time elapses. Every 5,760 years the carbon-14 decreases by 1/2 its value. This is a value which has been determined in the laboratory. If, say at the time of the death of a tree when the exchange between the tree and atmosphere ceases, there was a certain quantity of carbon-14 in the tree, after 5,760 years the quantity of carbon-14 would be only 1/2 its original value. An equation can be worked out relating the ratio of carbon-14 to carbon-12 in the organism at the time the sample is analyzed to the ratio of carbon-14 to carbon-12 that was present in the organism at the time of the cessation of the exchange with the atmosphere, the half-life of carbon-14, and the time that has elapsed since the exchange ceased. This equation can then be solved to give the alleged interval of time from the cessation of the exchange up to the present time. Actually the time of the analysis is not used, but rather the date 1950. This is before nuclear bomb testing artificially changed the C^{14} to C^{12} ratio.

The radioactivity of carbon-14 is very weak and even with all its dubious assumptions the method is not applicable to samples that supposedly go back 10,000 to 15,000 years. In those intervals of time the radioactivity from the carbon-14 would become so weak that it could not be measured with the best of instruments. Claims have been made that dating can be done back to from 40 to 70 thousand years. but it seems highly improbable that instruments could measure activity of the small amounts of C^{14} that would be present in a sample more than 15,000 years old.

2. Assumptions

Let us consider the assumptions involved in this method.

a. Variability of the concentration

If the concentration of C^{14} at the death of the organism being dated differed from the standard concentration, the date determined will not be meaningful unless we know what the concentration was. This would seem impossible to determine. Let us look at the agents which might affect this C^{14} concentration.

(1) Cosmic radiation bombardment

If there were less carbon-14 being produced in the past

due to weaker cosmic radiation bombardment than today, then, of course, the$(C^{14})_0$, representing the amount of carbon-14 at the time that the exchange ceased between the sample and the atmosphere, would be smaller. If the $(C^{14})_o$ were smaller than present-day concentration, then this would cause the apparent age to be somewhat larger than it should be. Until this point is settled, the value of the carbon-14 dating method is very questionable. Some investigators believe that the cosmic radiation bombardment from outer space has been constant or has not changed significantly. The cosmic radiation bombardment may have been smaller or may have been greater. Who knows, with any real certainty? A discussion of work by Prof. T.G. Barnes which follows indicates there has been considerable variation.

(2) **Atmospheric protection may have been provided by:**

(a) **A vapor canopy**

Many creationists believe that a vapor canopy covered the earth before the worldwide Flood. This would have been part of the source of the deluge released at the time of the Flood. This would also explain why people lived longer before the Flood; the earth was protected from the aging effect of cosmic rays by the absorbing action of the canopy. Furthermore, this canopy could have affected the C^{14} dating method because lowered cosmic ray incidence would have caused less C^{14} to be formed in the atmosphere.

(b) **Shielding by the magnetic field**

Recently, an article published by Professor Thomas Barnes of the University of Texas at El Paso shows that the main magnetic field of the earth is decreasing with time at a very rapid rate.[44] Professor Barnes assumes that the main magnetic field of the earth is generated by currents initiated in the fluid core of the earth at the time of the Creation. A flow of charged particles will induce a magnetic field. Most geophysicists believe that the main magnetic field is caused by circulating currents in the core. Now, these currents have energy associated with them. A charge flowing through a conductor, of course, gives up some of the energy it possesses as it moves through this conductor, producing heat. It does not seem that these currents in the core of the earth could be self-sustaining and last for aeons of time at the great strength they would have had to possess in times past. The currents would, with the passage of time, lose their energy, and the magnetic field would grow weaker due to this loss of energy.

Horace Lamb in the early 1880's made some calculations regarding the changing size of the earth's main magnetic field. He took Gauss' original measurements, which were made around 1835, and some measurements made after that, and found that the field was decaying with time. Since Lamb's work, many measurements have been made which support his findings.

If a curve of the strength of the field versus time is plotted, the manner in which this field is decaying can be determined. Dr. Barnes found the decay to be exponential. If the decay is exponential, the half-life for the magnetic field strength can be determined. Professor Barnes found that this half-life is about 1,400 years. This means that 1,400 years after it originated, the magnetic field strength would be down to 1/2 its original strength. At the end of 2,800 years, it would be down to 1/4 (1/2 of 1/2) of its original strength, and so on. After seven half-lives, or 9,800 years, it would have decayed to less than 1% of its original strength.

The magnetic field of the earth provides a shield against very high velocity particles that could cause great damage to living organisms. Should this magnetic field go down essentially to zero, there would be no protection against these high velocity particles for organisms here on the earth, regardless of the state of the atmosphere.

Further, there is energy associated with the flow of these currents in the core of the earth. They are not self-sustaining, and there is apparently no way to change the strength of the field. The energy associated with the flow of the currents in the core of the earth will be dissipated into the core of the earth in the form of heat energy. This heat energy will flow outward from the core to the other parts of the earth, changing the heat balance, or the heat budget, existing in the earth as time elapses.

If one works backward in time, taking the strength of

the field today (which is very well measured), the Joule heating in the past associated with the then existing currents in the core of the earth can be calculated. These calculations show that beyond 20,000 B.C. these currents and the resultant Joule heating would be so great that the heat would tend to separate the core and the mantle. If you go back even further, say, a million years, there would be so much heating associated with these currents that the whole earth would be destroyed by the heat energy.

The strength of the magnetic field thus indicates that the earth cannot be very old, and certainly Prof. Barnes' calculations indicate that 20,000 years is an absolute maximum for the age of the earth and that 10,000 years is a far more reasonable value.

Evidence that the magnetic field of the earth is rapidly decaying is of crucial importance to the carbon-14 dating method. The magnetic field strength of the earth very significantly affects the influx of cosmic radiation.

As already mentioned, cosmic radiation consists of high-speed protons moving through space. The magnetic field forms a barrier to the influx of cosmic radiation, and tends to channel what radiation does reach the earth into the polar regions. If the field were stronger in the past, it would be certain that the influx of cosmic radiation has been less than today. This, in turn, would mean that the production of neutrons in the atmosphere would have been less in the past. It would further mean that collisions of these neutrons with nitrogen atoms would have been fewer in number, and the production of carbon-14 in the past would thus have been less than it is today.

As a result, when a sample is taken from the stratosphere today, making allowances for the explosions of hydrogen and atom bombs and other events that might have put carbon-14 or carbon-12 into the atmosphere, the concentration of carbon-14 in this sample would exceed the concentration of C^{14} in the atmosphere, say 1,400 years ago, 2,800 years ago, or 4,200 years ago. This would affect $(C^{14})_0$ in the time equation very radically, and obviously the ages obtained for archaeological artifacts, for remains of men, for geological events associated with these archaeological artifacts and sites would be too large. This very rapid change in the magnetic field poses a very strong objection to the validity of the calculations in the carbon-14 dating method.

(3) **Nitrogen-14 content**

What was the N^{14} content in the atmosphere in the past? The supply of N^{14} would have an important effect on production of C^{14}. This poses a real uncertainty with no definite answer; therefore, it is very proper to question the assumptions made regarding the N^{14} content and consequent effect on the assumed C^{14} content of the earth's atmosphere in the past.

b. **The decay rate of C^{14} may have varied.**

In this dating method, as in the others, it is assumed that the decay rate is truly a constant. There is experimental evidence today that this is not so.

(1) **Electric fields**

The decay rate of Fe^{57} has been changed by as much as 3% with the use of electric fields. Thus it is possible that the decay rate of C^{14} and of other radioactive isotopes may have varied in the past due to unknown conditions.

(2) **Non-predictability of decay rates**

In a recent publication it has been reported that decay rates of C^{14} may not be predictable as previously believed.[45] Dr. John L. Anderson performed experiments on molecular mono-layers with C^{14} added. The emitted radiations did not occur in the pattern that classical theory assumes. Dr. Anderson's findings were confirmed through independent research at Atomic Energy Commission Laboratories. As with all new findings which run contrary to classical theory, there is skepticism, but there has as yet been no contradiction to Dr. Anderson's results. If his findings are confirmed by further study it will mean that decay rates are not fully predictable. Of course, as already mentioned, a constant decay rate, or at least a predictable decay rate, is the backbone of radiometric dating.

c. **Equilibrium**

If the earth is very old, the rate of production of C^{14} in the upper atmosphere, due to cosmic radiation as previously described, should be equal to the rate of decay. This equilibrium condition would have been reached when the rate of decay of C^{14} became equal to the rate of production of C^{14}. Libby and others who make use of the C^{14} dating method with great confidence believe that this equilibrium condition has been reached in the atmosphere of the earth, although Libby's first measurements showed that this was not true. It has been calculated that about 30,000 years are needed for the C^{14} to reach an equilibrium state.

If, as the evolutionists believe, the earth is billions of years old, the equilibrium state in the upper atmosphere would have been reached long ago. There is a very strong evidence, however, that this is not the case. Dr. Melvin A. Cook has developed serious objections to Libby's assumptions.[46]

It has been determined that the rate of formation of carbon-14 in the atmosphere is 2.5 atoms per square centimeter per second and the rate of decay of carbon-14 is 1.9 atoms per square centimeter per second. These two rates should be equal, if equilibrium exists, but there is a very significant difference. A difference of 0.6 carbon-14 atoms per square centimeter per second amounts to about 24%. How can an equilibrium condition be assumed with such a large difference existing?

There have been other values calculated for the rate of production and decay, and these indicate even larger percentage differences.[47]

A comparison of the equilibrium and non-equilibrium ages has been made be Dr. Cook.[48] Assuming a non-equilibrium state, a differential equation can be established with which the age of the earth can be calculated.[49] As the evidence indicates, it is assumed that the C^{14} is being produced at a faster rate than it is decaying.

This is an unbalanced condition very similar to the situation when water is pouring into a sink at a faster rate than it is going out. The interval of time it will take for the sink to fill up can be calculated very simply. The same holds true in regard to the C^{14} in the atmosphere of the earth. A similar calculation can be used to determine the length of time it would require for the atmosphere to gain the concentration of carbon-14 that it has today. This would also yield the age of the atmosphere. Taking the non-equilibrium model, Dr. Cook calculated an age of less than 10,000 years.[50] Certainly then, the earth and the inhabitants cannot be older than this. Dr. Cook believes that the "beginning" of radiocarbon in the atmosphere seems not to be at the time of creation of the earth, but rather at the time of the Flood, perhaps about 4,500 years ago. He believes that the whole atmosphere was fluxed so as to be cleansed of radiocarbon during the Flood. He believes further that most of the carbon in the antediluvian carbon cycle in the hydrosphere, the lithosphere, the atmosphere, and the biosphere were apparently locked out of the cycle by sudden deposition along with all but the representatives of the species taken into Noah's ark as were needed to regenerate the biosphere.

d. **No contamination is generally assumed.**

It is assumed that no seepage or other agent has added C^{14} to the specimen being dated since exchange has ceased. This assumption seems completely invalid. Bone has been found to absorb organic material, which contains carbon, from its surroundings. It is believed that enough C^{14} has been absorbed that its present content has no relationship to its original content of C^{14} and thus the apparent date becomes completely meaningless. There is no way to determine whether a sample is free of foreign carbon or not.

3. Summary of the Evidence Related to the Carbon-14 Dating Method

The C^{14} dating method is very questionable because of likely changes in the rate of production of C^{14} in the past and because of the unknown decay rates of the past. The very great possibility of contamination is another major difficulty. Finally, the use of an equilibrium model when the data show this condition does not exist eliminates any possibility of getting correct dates with this method.

Chapter 4
CONCLUSIONS

In surveying the various radiometric dating methods, we have seen that each is based on questionable assumptions. A major assumption regards the initial amounts of the parent and daughter elements present in the rocks. In some, hypotheses are made about the origin of the earth and its crust. In the uranium-thorium-lead method such a hypothesis is the basis of the guess as to the initial amounts of U and Th.

In the carbon-14 dating method it is assumed that the earth has existed long enough that carbon-14 must be in a state of equilibrium. This is assumed in spite of the fact that the measured decay and formation rates differ by at least 24%.

In the U-Th-Pb method, in addition to the assumptions concerning the initial amounts of U and Th, there are also assumptions concerning the Pb^{206}, Pb^{207}, Pb^{208}, and He^4 that may have been in the rocks originally and thus may not have been entirely the products of radioactive decay.

In the rubidium-strontium method, the initial concentration of Sr^{87} is not known, but Sr^{87} is at least 10 times more abundant than it should be if all of it were formed in 5 billion years by radioactive decay of Rb^{87}. Since these initial amounts are unknown, they can be "juggled" by geochronologists. Ratios of initial Rb^{87}/Rb^{86} and Sr^{87}/Sr^{86} can be adjusted so that any age is obtainable.

In the potassium-argon method, it is assumed that the initial Ar content of the rock is not important, though its abundance in the lithosphere and atmosphere is about 100 times the amount that would be generated by radioactive decay over 4.5 billion years.

"Juggling" is also performed by geochronologists in this K-Ar system. Here the actual observed branching ratio is not used, but rather a small ratio is arbitrarily chosen in an effort to match dates obtained by this method with U-Th-Pb dates.

The rates of formation of certain physical quantities must be known, but the possibility of an alteration of these rates in the past is not considered by the uniformitarians. For example, in the C^{14} method, C^{14} formation and, thus, its concentration, can be affected by changes in cosmic ray influx, atmospheric protection provided by a vapor canopy, a stronger magnetic field in the past, and a differing N^{14} content of the atmosphere.

The decay rates of the long-lived radioactive minerals themselves seem to have changed as evidenced by pleochroic halos. We know cosmic radiation can change decay rates and there is now evidence that decay rates are not as predictable as previously thought. If the decay rates have changes, the apparent ages obtained by radioactive dating methods have no relevance to the real ages.

Little attention is given to the leaching, contamination, and diffusion of elements involved while the decay processes occur. Helium, a product of uranium decay, escapes from rocks very readily, as does the argon associated with decay of potassium. These elements move upward through the crust and become concentrated at the surface where dating samples are being taken. Potassium and uranium are among the products which, in the form of water-soluble salts, can easily be leached out of rocks by water. Rubidium and strontium can also diffuse and become contaminated easily. It is admitted in this method of dating that individual grains of the same rock give different ages. This is because of diffusion. Radiochronologists will not admit that this could occur between rocks, however. Contamination can easily occur in the C^{14} method. We know argon can be inherited because pillow lavas (lavas which formed under water) known to be young have been dated as very ancient due to this phenomenon. Argon can also easily diffuse into a rock, more easily into some rocks than others.

The erratic dates obtained on moon rocks are examples of how contamination and diffusion can completely invalidate dates. Contamination in the physical sense is not the only way the picture can be misread. For example, in the U-Th-Pb dating methods neutron reactions seem to occur which "contaminate" the true decay products with isotopes which are identical, but are not a product of the decay.

When the dating methods are properly evaluated, we find that they give evidence for a short age of the earth. In the case of pleochroic halos, they show evidence that the original crust was formed very rapidly.

Neutron reaction corrections in the U-Th-Pb series reduce "ages" of billions of years to a few thousand years because most of the Pb can be attributed to neutron reactions rather than to radioactive decay.

On the basis of Dr. Barnes' study of the decay of the magnetic field, a ceiling may be placed on the age of the earth at 20,000 years. The strength of the magnetic field at this time would produce enough heat to separate the core and mantle. Going further back in time, eventually a time would be reached (at around one million years) when the earth would have vaporized due to the Joule heating of the currents causing the tremendously large field (3×10^{21} tesla)[9] that would exist then. The earth obviously cannot be that old.

Dr. Cook, taking the non-equilibrium model for C^{14} in the atmosphere of the earth, has found that it would take only about 10,000 years to gain the concentration of C^{14} that exists today, assuming none were present at the beginning.

With Dr. Cook's Q-expression, he has shown that either the lead isotope method of dating is invalid or the earth is so young that there have been no appreciable changes in the Q-factor to reflect a time-index.

In a recent article, Sidney P. Clementson[51] studied the application of radioactive dating methods to sedimentary rocks from a different approach than that of Dr. Cook. He investigates the basic assumption that radioactive disintegration starts when the minerals enter the host rocks and reaches the conclusion that this assumption is completely unreasonable. Clementson's calculations regarding young rocks leads him to conclude that radioactive disintegration in them is already at an apparent advanced age. Clementson concludes that "the theoretical ages calculated from isotope ratios are not the ages of the rocks, or of the earth, but are simply ratios of the minerals themselves which originated in the crust of the earth."[52]

An interesting hypothesis appeared recently regarding the supernova-type explosions that occur in space. Quoting the discussion directly: "The remnant of that local big bang is a pulsar called Vela-X (PSR 0833-45), which recent observations have positioned in the southern sky some 1,500 light years away, and which is considered to have given rise to the huge Gum Nebula. . . Being so close, the anisotropic neutrino flux of the super-explosion must have had the peculiar characteristic of resetting all our atomic clocks. This would knock our carbon-14, potassium-argon, and uranium-lead dating measurements into a cocked hat!"[53] It must be stressed that this conclusion was based on speculative ideas about the possible effect of neutrinos on radioactive decay. Nevertheless, it does emphasize that radioactive dating methods are based on the assumption of uniform conditions of the past, and this assumption may be totally invalid.

If the earth-moon system is as old as alleged by the evolutionists, there should be no residual Neptunium-237, uranium-236, and plutonium-244.

Radioactive minerals with "small" half-lives are being found by investigators. Neptunium-237 ($T_{1/2}$ = 2.2 million years) and uranium-236 ($T_{1/2}$ = 24 million years) have been discovered in moon rock samples.[54] Plutonium-244 ($T_{1/2}$ = 76 million years) has been found in earth rocks.[55]

Strong credence is given to Dudley's theory by the recent experiments that seem to show that neutrinos come in three varieties.[56] These neutrinos can move at various speeds and seem to have a small rest mass.

The writer heard a man who was associated with the radiocarbon dating laboratory of the University of Pennsylvania state that dates that did not conform to the evolutionary concept of the age of the earth and geological chronology were discarded as erroneous. The evolutionist has thus imposed upon science his own philosophy regarding the age and origin of things. We have heard astronauts on the moon cry with great excitement that they see a rock at the foot of a mountain that is at least 3 billion years old. Amazing! What discernment they possess if they can just look at a rock and tell us its age. Here is obvious evidence that the thing which was to be proved has been assumed.

It appears that geochronology is an example of a mixture of assumptions, guesses, and imposed imagined universal principles. It would seem that the case of modern geochronology is established in the manner of Mr. Enlightenment's statement in C.S. Lewis' *Pilgrim's Regress:* "Hypothesis, my dear young friend, establishes itself by a cumulative process: or, to use popular language, if you make the same guess often enough it ceases to be a guess and becomes a Scientific Fact. . . But when you have had a scientific training you will find that you can be quite certain about all sorts of things which now seem to you only probable."

REFERENCES

1. The following papers show the views held by the advocates of two different camps regarding the nucleogenesis of elements in the earth's crust:
   Gerling, E.K., "Age of the Earth According to Radioactivity Data," *Compt. Rend. Acad. Sci. U.R.S.S. 34,* 1942, pp. 259-261;
   Houtermans, F.G., "Isotopenhaufigkeiten Im Naturlichen Blei und das Alter des Urans," *Naturwissenschaften 33,* 1946, pp. 185-186;
   Houtermans, F.G., "Das Alter des Urans," Z. *Naturforsch., IIa,* pp. 322-328;
   Holmes, Arthur, "Estimate of the Age of the Earth," *Nature 159,* 1946, pp. 127-128;
   Jeffreys, H., "Lead Isotopes and the Age of the Earth," *Nature 162,* 1948, pp. 822-823;
   Bullard, E.C., and Stanley, J.P., "The Age of the Earth," *Suomen Geodeettisen Laitoksen, Julkaisuja: Verofentl. Finnisch. Geodat. Inst.,* No. 36:33-40, 1949;
   Alpher, R.A., and Herman, R.C., "The Primeval Lead Isotopic Abundances and the Age of Earth's Crust," *Phys. Rev. 84,* 1951, pp. 1111-1114.
2. Cloud, Preston (Ed.), *Adventures in Earth History,* San Francisco: W.H. Freeman and Co., 1970, p. 101.
3. Holmes, Arthur, *Principles of Physical Geology,* New York: Ronald Press, 1965, p. 346.
4. Ibid., p. 346.
5. Faul, Henry, *Ages of Rocks, Planets, and Stars,* New York: McGraw-Hill, Inc., 1966, p. 2.
6. Ibid., p. 3.
7. Dudley, H.D., *The Morality of Nuclear Planning,* Glassboro, 1976, p. 52.
8. Ibid., p. 53.
9. Ibid., p. 54.
10. Ibid., p. 55.
11. Dudley, H.C., *Phenomenological Causal Model of Nuclear Decay, Assuming Interaction with Neutrino Sea, Lettere Al Nuovo, Cimento,* Vol. 5, No. 3, September 16, 1972, p. 232.
12. Dudley, H.C., *The Morality of Nuclear Planning,* Glassboro, 1976, pp. 56-57.
13. Lee, T.D., and Yang, C.N., *Phys. Rev. Lett.,* 4, 1960, p. 307.
14. Faul, Henry, *Ages of Rocks, Planets, and Stars,* New York: McGraw-Hill, Inc., 1966, pp. 40-41.
15. Knopf, Alfred (Ed.), *Age of the Earth,* Bull. 80, National Research Council, 1931, p. 107.
16. Allen, Roy M., "The Evaluation of Radioactive Evidence on the Age of the Earth," *Journal of the American Scientific Affiliation,* December, 1952, p. 18.
17. Gentry, R.V., "Cosmological Implications of Extinct Radioactivity from Pleochroic Halos," *Creation Research Society Quarterly,* Vol. 3, No. 2, 1966, pp. 17-20.
18. Ibid., p. 19.
19. Anderson, John Lynde, and Spangler, George W., "Radiometric Dating: Is the "Decay Constant" Constant?", *Pensee,* p. 31.
20. Ibid., p. 33.
21. Faul, Henry, pp. 65-67.
22. Holland, G.D., and Kulp, J.L., *Geochim, et Cosmochim, Acta 5,* 197, 1945, p. 214.
23. Abranches, M.C.B., Arden, J.W., and Gale, N.H., "Uranium-Lead Abundances and Isotopic Studies in the Chondrites Richardton and Farmington Earth and Planetary Science Letters," 46, 1980, pp. 311-322.

24. Ibid., p. 317.
25. Ibid., p. 320.
26. Cook, Melvin A., Utah Engineering Experiment Station Bulletin 74, No. 16, 1955.
27. Ibid., p. 9.
28. Cook, Melvin A., *Prehistory and Earth Models*, London: Max Parrish and Co. Ltd., 1966, p. 62.
29. Cook, Melvin A., Utah Engineering Experiment Station Bulletin 74, No. 16, 1955, p. 18.
30. Ibid., Appendix II, p. iv.
31. Ibid., Appendix II, p. v.
32. Cook, Melvin A., and Cook, M. Garfield, *Science and Mormonism*, Deseret Books, p. 182.
33. *Chemical and Engineering News*, Jan. 17, 1972, p. 9.
34. Cook, *Prehistory and Earth Models*, p. 66.
35. Ibid., p. 67.
36. There are two papers of prominence reporting this and another case: *Journal of Geophysical Research*, 17, 1968, pp. 4601-4607; *Science, 162*, 1968, pp. 265-266.
37. Cook, *Prehistory and Earth Models*, 1966, p. 65.
38. From private discussion.
39. "Mantle Geochemistry: Probing the Source of the Earth's Crust," *Science*, Vol. 203, February 9, 1979.
40. Ibid., p. 530.
41. Brooks, C., James, D.E., and Hart, S.R., "Ancient Lithosphere: Its Role in Young Continental Volcanism," *Science*, Vol. 193, September 17, 1976, p. 1086.
42. Ibid., p. 1087.
43. Ibid.
44. Barnes, T.G., "Decay of the Earth's Magnetic Moment and the Geochronological Implications," *Creation Research Society Annual*, June, 1971, pp. 24-29.
45. Anderson, J.L., Abstract of Papers, 161st National Meeting, American Chemical Society, Los Angeles, 1971.
46. Cook, Melvin A., *Prehistory and Earth Models* p. 3.
47. Ibid., p. 5.
48. Ibid., p. 9.
49. Cook, Melvin A., "Geological Chronometry," Utah Engineering Station Bulletin 83, 47, No. 18, 1956, p. 8.
50. Cook, Melvin A., "Carbon-14 and the Age of the Atmosphere," *Creation Research Society Annual*, June, 1970, pp. 53-56.
51. Clementson, Sidney P., "A Critical Examination of Radioactive Dating of Rocks," *Creation Research Society Quarterly*, December, 1970, pp. 137-144.
52. Ibid., p. 137.
53. Jueneman, F.B., *Industrial Research*, Sept., 1972, p. 15.
54. *Chemical and Engineering News*, Jan. 17, 1972, p. 9.
55. *United Press Report*, Sept., 1971.
56. "Do Neutrinos Oscillate from One Variety to Another?" *Physics Today*, Vol. 33, July, 1980, pp. 17-19.